Penguin Education
Penguin Library of Physical Sciences

Orbitals and Symmetry
D. S. Urch

Advisory Editor
V. S. Griffiths

General Editors
Physics: N. Feather, F.R.S.
Physical chemistry: W. H. Lee
Inorganic chemistry: A. K. Holliday
Organic chemistry: G. H. Williams

Orbitals and Symmetry

D. S. Urch

Penguin Books

Penguin Books Ltd, Harmondsworth,
Middlesex, England
Penguin Books Inc., 7110 Ambassador Road,
Baltimore, Md 21207, U.S.A.
Penguin Books Australia Ltd, Ringwood,
Victoria, Australia

First published 1970
Copyright © D. S. Urch, 1970

Filmset in 'Monophoto' Times New Roman by
Keyspools Ltd, Golborne,
and made and printed in Great Britain by
Fleming & Humphreys (Baylis) Ltd

To Patricia

Contents

Editorial Foreword   9
Preface 10

Chapter 1 Atoms 11
1.1 Wave mechanics 11
1.2 The hydrogen atom 14
1.3 The helium atom 18
1.4 Energy states in the helium atom 21
1.5 The correlation of electrons 25
    Problems 1 28

Chapter 2 Molecules 29
2.1 The chemical bond 29
2.2 Examples of simple molecular orbital calculations 33
2.3 Variations of α or β 46
2.4 Electron repulsion in molecular orbitals 49
2.5 Molecular wave functions 51
2.6 Valence bond method 52
2.7 Configuration interaction 52
2.8 Conditions for bonding 53
    Problems 2 56

Chapter 3 Symmetry and Point Groups 57
3.1 Rotations 58
3.2 Reflections 63
3.3 Improper rotations 66
3.4 Inversion through a centre of symmetry 69
3.5 Multiplication of symmetry operators 70
3.6 Group theory 72
3.7 Classes of symmetry operators 73
3.8 Simple point groups 74
3.9 Other groups 78
3.10 Tetrahedral groups 81
3.11 Cubic or octahedral groups 83
3.12 Icosahedral groups 84
3.13 To determine the point group of a molecule 86
     Problems 3 87

Chapter 4 Character Tables 88
4.1 Reduction of matrices 94
4.2 The character 97
4.3 Character tables for point groups 98
4.4 Direct product 98
4.5 Reduction of large matrices using character tables 99
4.6 Irreducible-representation wave functions 102
4.7 Molecules with no central atom 104
4.8 Representations of molecular orbitals 105
    Problems 4 107

Chapter 5 Chemical Bonding 108
5.1 Hybridization 111
5.2 Promotion energy and resonance structures 116
5.3 Equivalent orbitals 117
5.4 Chemical applications 123
5.5 π-Bonding in organic molecules 130
5.6 'Electron-deficient' molecules 141
5.7 Resolution of a structural problem 148
5.8 Molecular shapes 150
5.9 Symmetry conservation in chemical reactions 155
    Problems 5 158

Chapter 6 Visible and Ultraviolet Spectroscopy 159
6.1 Selection rules 159
6.2 Chromophores 164
6.3 Crystal field theory 165
6.4 Spin–orbit coupling – double groups 177
    Problem 6 180

Chapter 7 Molecular Vibrations and Crystal Symmetry 181
7.1 Infrared and Raman spectroscopy 181
7.2 Selection rules for infrared and Raman spectra 186
7.3 Characteristic bond frequencies 190
7.4 Crystallography 192
7.5 The structure of crystals 194
7.6 Space groups 200
    Problem 7 203

Bibliography 205
Appendix A Matrix Multiplication 209
Appendix B Character Tables for Point Groups 211
Answers 237
Index 249

Editorial Foreword

The chemistry section of the *Penguin Library of Physical Sciences* is planned to cover the normal content of honours degree courses of British universities, and those of comparable standard – C.N.A.A, Royal Institute of Chemistry, etc. Most of the optional subjects, which are becoming a prominent feature of present-day courses, will be included.

The series has been planned as a whole: nomenclature, the enumeration of tables and diagrams, and their layout, have been standardized throughout, so that there is a minimum of duplication. Every student will agree that an appreciable part of almost any modern textbook of organic or inorganic chemistry includes an account of atomic structure, wave mechanics, atomic and molecular orbitals, hybridization, etc., before making use of these concepts. This kind of duplication may be eliminated, in a uniform series, by frequent cross referencing.

The subject has been subdivided so that students may buy the appropriate volumes as and when they become relevant to their course of study. Due to the introduction of optional subjects, and the advent of inter-disciplinary courses, not every student will require every volume. The outlay required, in any one year of the course, becomes realistic in present-day terms.

W.H.L.

Preface

Quantum-mechanical calculations and symmetry concepts are of increasing importance in chemistry. It is therefore reasonable that such topics should figure in the undergraduate teaching of chemistry. Two courses are open; a rigorous mathematical treatment of quantum chemistry or a less rigorous approach in which approximations are often made without adequate justification and many concepts are used that are not fully explained. The latter approach may be justified by claiming that results of chemical interest are more quickly and more easily obtained. The danger is that the results will be reduced in value because of the approximations they contain. While the first course holds out the hope of completely *a priori* calculations, the latter more approximate approach must make frequent recourse to experiment for comparison. Even so, simple and approximate calculations can often be of great value in understanding many chemical phenomena, and in suggesting new lines for research.

This book is non-rigorous. No attempt is made to deal in mathematical depth either with quantum mechanics or with group theory. However an attempt has been made to describe the limitations imposed by approximate methods, so that the reader may have a critical appreciation of the worth of the calculations and also to prepare for more sophisticated types of calculation. It is hoped in this way that the reader will acquire, for example, the ability to do simple Hückel calculations and also to realize their limitations.

This book has developed from lectures given to second- and third-year undergraduates in the chemistry department at Queen Mary College and a level of mathematical ability, roughly equivalent to 'A'-level of the G.C.E., has been assumed.

Chapter 1
Atoms

Nature is atomic. Atoms are the building bricks of matter. There are at the present time 104 chemically distinct types of atom known and under normal conditions it is not possible to change one atom into another. This process can be achieved by nuclear reactions, which also give us insight into the structure of atoms. Most of the mass and all the positive charge resides in the nucleus which is composed of positively charged protons and neutral neutrons. Protons and neutrons have approximately the same mass, about 1.66×10^{-24}g. In a neutral atom the number of negatively charged electrons around the nucleus will equal the number of protons in the nucleus. This is the *atomic number, Z*. Chemistry is concerned with interatomic interactions and therefore with interelectronic interactions, so the particular number of electrons (the same as Z) will define the chemical character of an atom. Thus all atoms with the same value of Z will be atoms of the same *element*. A particular number of protons may, however, be associated with different numbers of neutrons to yield nuclei of varying masses. Such nuclei are *isotopes*. The number, energy and spatial distribution of the electrons will be of paramount importance in chemistry. Sometimes electrons may be lost or gained, giving rise to positively or negatively charged ions respectively. Such ions then attract each other by electrostatic forces, forming ionic structures. Atoms can also share electrons; these electrons then move under the attraction of two nuclei rather than one and a covalent bond has been formed.

1.1 Wave mechanics

It is not possible to extend physical ideas that work well in the macroscopic world into the region of subatomic particles without some modification. It is necessary to admit that on occasions a particle may best be described by an equation that has the same form as a wave equation in classical physics. This is often spoken of as a 'duality' between particle and wave behaviour but it is as well to remember that the duality exists only in our minds and not in the nature of the electron. The problem arises because we are trying to describe electrons and protons with concepts that work better for buses and billiard balls.

In dealing with the behaviour of electrons in atoms and molecules we shall concentrate upon their wave nature and take as a fundamental equation

$$\frac{\partial^2 \psi}{\partial x^2} = -\frac{4\pi^2}{\lambda^2} \psi,$$

where ψ is the *wave function* – a function of x and time that will satisfy this equation – and λ is the wavelength. But if we now consider the particle nature of electrons, we can speak of mass m, velocity v, momentum p ($p = mv$) and kinetic energy T,

$$T = \tfrac{1}{2}mv^2 = \frac{p^2}{2m}.$$

An important factor that concerns subatomic particles is the quantization of energy. Quantization means that under many circumstances particles such as electrons or protons can only have particular energies. This is found for example in atoms. The electrons that surround the nucleus have only certain very well defined energies as a result of quantization.

De Broglie suggested a relationship that links the 'particle' and 'wave' concepts,

$$\lambda = \frac{h}{p},$$

where h is Planck's constant.

Planck's constant was originally postulated in the equation that links the energy of radiation E with its frequency v

$$E = hv.$$

Using de Broglie's relationship it is now possible to express the classical wave equation as

$$\frac{\partial^2 \psi}{\partial x^2} = -\frac{4\pi^2 p^2}{h^2}\psi,$$

where ψ is still a function of x and time.

Now the total energy E of a classical particle is the sum of its kinetic (T) and potential (V) energies, thus $E = T + V$. Since $T = p^2/2m$,

$$\frac{\partial^2 \psi}{\partial x^2} = -\frac{8\pi^2 m(E - V)}{h^2}\psi.$$

This may be generalized for three dimensions (i.e. ψ is now a function of x, y and z as well as time) and rearranged:

$$\left(\frac{\partial^2}{\partial x^2} + \frac{\partial^2}{\partial y^2} + \frac{\partial^2}{\partial z^2}\right)\psi + \frac{8\pi^2 m(E - V)}{h^2}\psi = 0,$$

which is usually abbreviated to

$$\nabla^2 \psi + \frac{8\pi^2 m(E - V)}{h^2}\psi = 0.$$

This is Schrödinger's equation for three dimensions and is of fundamental importance. Notice that it can also be written in the form

$$\left(-\frac{h^2}{8\pi^2 m}\nabla^2 + V\right)\psi = E\psi.$$

Within the brackets are a collection of mathematical operators which when they act upon the wave function ψ yield a scalar number, the total energy, multiplied by the wave function once again. When it is possible to arrange an equation in this way,

(Operator) function = (scalar number) × function,

then the function is said to be an *eigenfunction* of the operator and the scalar number is an *eigenvalue*. Before we return to the Schrödinger equation let us consider a few simple examples:

Operator   *Function*

$\dfrac{d}{dx}$ x^2 Not an eigenfunction since $\dfrac{d}{dx}x^2 = 2x$ (i.e. the same function does not occur on both sides of the equation).

$\dfrac{d}{dx}$ e^{ax} An eigenfunction with an eigenvalue, a:

$$\frac{d}{dx}e^{ax} = ae^{ax}.$$

$\dfrac{d}{dx}$ $\cos x$ Not an eigenfunction since $\dfrac{d}{dx}\cos x = -\sin x$.

$\dfrac{d^2}{dx^2}$ $\left\{\begin{array}{c}\sin x \\ \cos x\end{array}\right\}$ Both eigenfunctions with an eigenvalue, -1:

$$\frac{d^2}{dx^2}\sin x = -\sin x,$$

$$\frac{d^2}{dx^2}\cos x = -\cos x.$$

When the Schrödinger equation is written in this form the following symbolism is used,

$$\mathscr{H}\psi = E\psi.$$

The wave function of the system ψ is an eigenfunction of the operator $\mathscr{H}$ and the corresponding eigenvalue E is the total energy of the system.

The operator $\{-(h^2/8\pi^2 m)\nabla^2 + V\}$, to which the symbol $\mathscr{H}$ is given, is referred to as the Hamiltonian operator by analogy with equations in classical mechanics for total energy, derived by Hamilton.

The wave function ψ is a description of the distribution of the 'particle' in space. It is most convenient if such functions refer to just one particle, for example to just one electron. This would mean that the total probability of finding an electron should be just one. Now the probability of finding an electron described by a function ψ at any point in space is proportional to ψ^2. (If the function contains complex parts, such as $\psi = x + iy$, the probability is proportional to $\psi\psi^*$, where ψ^* is the complex conjugate, $\psi^* = x - iy$. This ensures that the probability is always real since $\psi\psi^* = x^2 + y^2$.) The summation of these probabilities at various points in space may be achieved by integration over the whole of space, thus $\int \psi^2 \, dv$ (where dv is an infinitesimally small volume element). This integral is to be made equal to $+1$, and when it is, the function ψ is said to be *normalized*. If the integral is not 1, but equal to, say A^2, then the normalized wave function would be $A^{-1}\psi$. (See problems 1.1 and 1.2.)

1.2 The hydrogen atom

In the hydrogen atom a negative electron moves under the attractive influence of a positive proton. The electrostatic potential is $-e^2/r$ so that the Schrödinger equation for the hydrogen atom is

$$\left(-\frac{h^2}{8\pi^2 m}\nabla^2 - \frac{e^2}{r}\right)\psi = E\psi.$$

The solution of the Schrödinger equation for the hydrogen atom is dealt with in standard texts and only a few of the wave functions will be given here (Table 1).

Table 1 Wave Functions for the Hydrogen Atom

$1s \qquad \psi = (a_0^3 \pi)^{-\frac{1}{2}} \exp -\dfrac{r}{a_0}$

$2s \qquad \psi = (32 a_0^3 \pi)^{-\frac{1}{2}}\left(2 - \dfrac{r}{a_0}\ \exp -\dfrac{r}{2a_0}\right)$

$2p_z \qquad \psi = (32 a_0^3 \pi)^{-\frac{1}{2}}\dfrac{r}{a_0}\cos\theta \exp -\dfrac{r}{2a_0}$

$2p_x, 2p_y \quad \psi = (64 a_0^3 \pi)^{-\frac{1}{2}}\dfrac{r}{a_0}\sin\theta \exp -\dfrac{r}{2a_0}\exp \pm i\phi$

By taking suitable linear combinations of the $2p_x$, $2p_y$ orbitals the following real orbitals are obtained:

$2p_y \qquad \psi = (32 a_0^3 \pi)^{-\frac{1}{2}}\dfrac{r}{a_0}\sin\theta \sin\phi \exp -\dfrac{r}{2a_0}$

$2p_x \qquad \psi = (32 a_0^3 \pi)^{-\frac{1}{2}}\dfrac{r}{a_0}\sin\theta \cos\phi \exp -\dfrac{r}{2a_0}$

$$a_0 = \frac{h^2}{4\pi^2 m e^2} = 0.0529 \text{ nm}$$

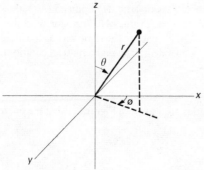

Figure 1   Coordinates of point (x, y, z) equivalent to (r, θ, ϕ)

These wave functions are most conveniently expressed in polar, not Cartesian, coordinates. The relationship between these two coordinate systems is shown in Figure 1. The permitted wave functions for the hydrogen atom are governed by three *quantum numbers*: principal n, angular momentum l and magnetic m_l. Only certain integral values are allowed; for n: 1, 2, 3, ... etc.; for l: 0, 1, 2, ..., n; for m_l: 0, ± 1, ± 2, ..., $\pm l$. The energies of the wave functions are, in the absence of any other effects, governed only by the principal quantum number. It is found that it is possible to express the wave functions as the product of two functions, one dependent only on the angular coordinates θ and ϕ and one dependent only on r. Further it is found that the former, angular, functions contain only the quantum numbers l and m_l whereas the latter r-functions depend on quantum numbers n and l. The shapes of orbitals are controlled by the angular functions and therefore upon l and m_l. Numerical values for l are usually replaced by letters thus: $l = 0, s$; $l = 1, p$; $l = 2, d$; $l = 3, f$; $l = 4, g$; etc. Associated with each value of l will be $2l+1$ orbitals differentiated only by values of m_l. Thus for p-orbitals, $m_l = +1, 0, -1$; for d-orbitals, $m_l = +2, +1, 0, -1, -2$; and so on. As can be seen from Table 1 the wave functions involve real and complex parts, but it is

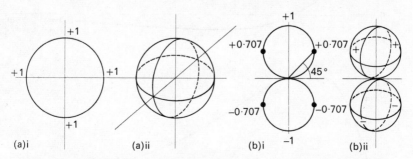

(a)i   (a)ii   (b)i   (b)ii

Figure 2   Angular parts of real wave functions. (a) $l = 0$, s-orbital: i, variation of θ in a plane of constant ϕ; ii, variation of θ and ϕ in space. (b) $l = +1$ ($m_l = 0$), p_z orbital: i, variation of θ in a plane of constant ϕ: ii, variation of θ and ϕ in space

possible by taking suitable linear combinations of orbitals of different m_l values to construct $2l+1$ real *orthogonal* orbitals for each value of l. Orthogonal orbitals are those which are completely independent in space in the same way as the three Cartesian coordinates are completely independent. The mathematical condition for orthogonality between ϕ_a and ϕ_b is

$\int \phi_a \phi_b \, dv = 0.$

The angular parts of real wave functions are shown in Figure 2, using the polar coordinate representation.

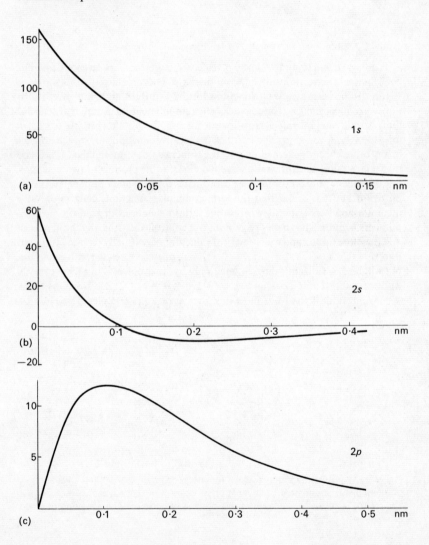

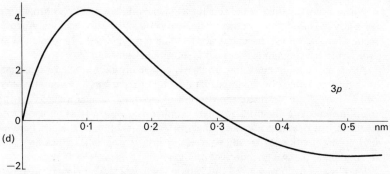

(d)

Figure 3   Radial wave functions for various values of n and l: (a) 1s (b) 2s
(c) 2p (d) 3p. The vertical axes represent the radial part of the wave function in
units of nm $^{-\frac{3}{2}}$

These figures do not of course give a true indication of electron distribution
in an orbital since they only represent part of a wave function. Each of the angular
functions must be multiplied by the radial part of the wave function. Some typical
radial functions for various values of n and l are shown in Figure 3. As can be
seen when $l = n-1$ the function does not cut the radial axis and does not change
sign. However when $l = n-2$ the function does change sign at a particular
distance. This distance is thus the radius of a sphere where the whole wave
function will be zero: a spherical node. The number of such nodes is $n-l+1$.

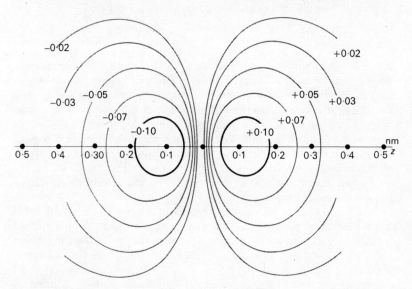

Figure 4   Cross-section through the wave function for a hydrogen $2p_z$ orbital in
a plane containing the z-axis

17   The Hydrogen Atom

When trying to visualize the actual wave function for a particular orbital it is necessary to consider the product of the two parts of the wave function. A cross-section through such a product wave function for a hydrogen $2p_z$ orbital in an arbitrary plane containing the z-axis is shown in Figure 4.

A further question must now be asked: how many electrons can we put into each orbital, defined by the three quantum numbers n, l, m_l? The question may be answered if we postulate the concept of *electron spin* and assume that an electron can rotate about its own axis in either a clockwise or an anticlockwise direction. The quantum number that arises from a quantum-mechanical description of this 'motion' is m_s and it can have just two possible values, $\pm\frac{1}{2}$, corresponding to the two directions of rotation.

Since the electron is charged this rotation will generate a magnetic moment; that is, a spinning electron behaves like a magnet. If a magnetic field is applied the 'electron magnet' can take up two possible orientations depending on the direction of the electron spin. The component of the moment in the field direction has a value

$$\frac{1}{2}\frac{e}{m}\frac{h}{2\pi} = \frac{eh}{4\pi m} = 1 \text{ Bohr magneton} = 9\cdot27\times10^{-24}\ \text{A m}^2\,(\text{J T}^{-1}).$$

If we further assume that each electron in an atom must be defined by a discrete and independent set of four quantum numbers (the Pauli exclusion principle), then it follows that the maximum number of electrons that could ever be found in one orbital is two. For these two electrons n, l and m_l would be the same but they would have different values of m_s: one $+\frac{1}{2}$, the other $-\frac{1}{2}$.

1.3 The helium atom

The most tightly bound orbital in helium is the $1s$. If this houses both electrons this would be the most stable arrangement of electrons (ignoring electron repulsion) and is written $1s^2$. The allocation of electrons to orbitals specifying n and l is called an *electronic configuration*. The first electronically excited configuration would correspond to one of the electrons being placed in an orbital of principal quantum number 2: $1s^1 2s^1$, or $1s^1 2p^1$. In the absence of a magnetic field, orbitals differentiated only by m_l values have the same energy, for example the three $2p$ orbitals with $m_l = +1$, 0 and -1. When two or more wave functions have the same energy they are said to be *degenerate*; the three $2p$ orbitals provide an example of threefold degeneracy.

The effect of electron repulsion and the added complication of electron spin do however remove the degeneracy associated with l (for a given value of n) found in hydrogen. Let us first try and evaluate some of the energies for the configuration of the helium atom mentioned above to demonstrate the general effect of electron repulsion.

Each electron carries a negative charge and the Hamiltonian operator to evaluate the energy of the helium atom will have to take electron repulsion into account. The contribution to the potential energy will be just one term of the type $+e^2/r(12)$ where $r(12)$ is the distance between the two electrons 1 and 2. With the inclusion of this term in the Hamiltonian the hydrogen-like wave functions will no longer be eigenfunctions of the operator. In this case it is necessary to evaluate an average value of the energy by considering the energy at each point in space and summing (i.e. integrating) over the whole of space:

$$E = \frac{\int \psi \mathcal{H} \psi^* \, d\tau}{\int \psi\psi^* \, d\tau}.$$

Here ψ is a function of both space and spin coordinates and $d\tau$ is an infinitesimally small element in the same set of coordinates.

We must also pay particular attention to the nature of the wave function we use in this equation. It is tempting to try to use the orbitals, that is the wave functions, that have been used for just one electron moving around a nucleus of charge $+1$ for the two electrons of the helium atom, moving now round a charge of $+2$. Clearly the potential field which each electron feels due to the nucleus can be easily evaluated, it will be just $-Z(e^2/r)$ for each electron ($Z = 2$ for helium). However in the helium atom this wave function must describe the behaviour of both electrons. The simplest type of function would assume that both electrons are independent of each other and so simple one-electron functions could be run together in some way to generate an over-all wave function. If these two one-electron functions were $\psi(1)$ and $\psi(2)$ their possible over-all functions might be

(i) $\Psi = \psi(1) + \psi(2)$

or (ii) $\Psi = \psi(1)\psi(2)$.

The Hamiltonian operator has so far been developed for just one electron, say $\mathcal{H}(1)$. Clearly the operator for many electrons would be the sum of the individual operators, thus for helium $\mathcal{H}(\text{total}) = \mathcal{H}(1) + \mathcal{H}(2)$. Also the Hamiltonian for a particular electron will only act upon a wave function for that electron, e.g. $\mathcal{H}(1)\psi(1) = E_1\psi(1)$ but $\mathcal{H}(1)\psi(2) = 0$. Let us now test the functions (i) and (ii) above,

$$\mathcal{H}(\text{total})\Psi_{(i)} = [\mathcal{H}(1) + \mathcal{H}(2)][\psi(1) + \psi(2)]$$
$$= E_1\psi(1) + E_2\psi(2),$$

but $\mathcal{H}(\text{total})\Psi_{(ii)} = [\mathcal{H}(1) + \mathcal{H}(2)][\psi(1)\psi(2)]$
$$= E_1\psi(1)\psi(2) + \psi(1)E_2\psi(2)$$
$$= (E_1 + E_2)[\psi(1)\psi(2)]$$
$$= E(\text{total})\Psi_{(ii)}.$$

Only the product wave function (ii) gives equations of the right form. There is one fundamental condition which these wave functions must obey; that the over-all wave function must be antisymmetric with respect to the interchange of any pair of electrons. This means that if any two electrons are allocated to each other's orbitals then the over-all wave function will change sign. The simplest way this can be achieved in the general case is by use of a determinant and allocating one electron to each row or column. This is because the interchange of a row or column will have the same effect as multiplying the determinant by -1. The wave functions to be used for individual electrons will have to specify not only orbital (i.e. spatial distribution) but also the spin function to be used. If ϕ_1, ϕ_2, etc. are typical orbitals then each may be associated with spin $m_s = +\frac{1}{2}$ or $m_s = -\frac{1}{2}$ usually symbolized α or β. A general wave function would then be

$$\Psi = \frac{1}{\sqrt{(n!)}} \begin{vmatrix} \psi_1\alpha(1) & \psi_1\beta(1) & \psi_2\alpha(1) & \psi_2\beta(1) & \dots & \psi_{\frac{1}{2}n}\alpha(1) & \psi_{\frac{1}{2}n}\beta(1) \\ \psi_1\alpha(2) & \psi_1\beta(2) & \psi_2\alpha(2) & \psi_2\beta(2) & \dots & \psi_{\frac{1}{2}n}\alpha(2) & \psi_{\frac{1}{2}n}\beta(2) \\ \psi_1\alpha(3) & \psi_1\beta(3) & \psi_2\alpha(3) & \psi_2\beta(3) & \dots & \psi_{\frac{1}{2}n}\alpha(3) & \psi_{\frac{1}{2}n}\beta(3) \\ \vdots & \vdots & \vdots & \vdots & & \vdots & \vdots \\ \psi_1\alpha(n) & \psi_1\beta(n) & \psi_2\alpha(n) & \psi_2\beta(n) & \dots & \psi_{\frac{1}{2}n}\alpha(n) & \psi_{\frac{1}{2}n}\beta(n) \end{vmatrix},$$

and is called a *Slater determinant*.

If n is the number of electrons involved then $(n!)^{-\frac{1}{2}}$ is the normalizing factor to take account of the $n!$ terms that will result from the expansion of the determinant.

For the ground state of helium the electronic configuration is $1s^2$. The corresponding over-all wave function will be

$$\Psi = \frac{1}{\sqrt{2}} \begin{vmatrix} 1s\alpha(1) & 1s\beta(1) \\ 1s\alpha(2) & 1s\beta(2) \end{vmatrix}, \qquad \textbf{1.1}$$

which on expansion yields

$$\Psi = \frac{1}{\sqrt{2}}[1s\alpha(1)1s\beta(2) - 1s\beta(1)1s\alpha(2)].$$

The orbital functions may be factorized out,

$$\Psi = 2^{-\frac{1}{2}}.1s(1)1s(2)[\alpha(1)\beta(2) - \beta(1)\alpha(2)].$$

If the electrons are interchanged, equation **1.1** becomes

$$\Psi' = 2^{-\frac{1}{2}}1s(2)1s(1)[\alpha(2)\beta(1) - \beta(2)\alpha(1)]$$
$$= -2^{-\frac{1}{2}}1s(2)1s(1)[\beta(2)\alpha(1) - \alpha(2)\beta(1)].$$

This equation only differs from $-\Psi$ by the order in which the functions for electrons 1 and 2 are written down. Clearly this will not affect the value or sign of $-\Psi$, therefore $\Psi' = -\Psi$ as required above. Notice that in this example the orbital function is symmetric with respect to electron interchange and the spin function is antisymmetric. The division into spin and orbital functions is often very useful but for their product to be antisymmetric, symmetric spin functions can only be aligned with antisymmetric orbital functions and vice versa. Only a symmetric orbital function can be constructed for the configuration $1s^2$ but for the first excited state, $1s^12s^1$, the following may be considered (symmetric and antisymmetric functions are distinguished by $(+)$ and $(-)$):

$$\psi_{orb}(+) = 2^{-\frac{1}{2}}[1s(1)2s(2) + 2s(1)1s(2)], \qquad \textbf{1.2}$$

$$\psi_{orb}(-) = 2^{-\frac{1}{2}}[1s(1)2s(2) - 2s(1)1s(2)]. \qquad \textbf{1.3}$$

Corresponding spin functions are

$$\psi_{spin}(+)_{(i)} = 2^{-\frac{1}{2}}[\alpha(1)\beta(2) + \beta(1)\alpha(2)] \qquad \textbf{1.4}$$

and $\quad \psi_{spin}(-) = 2^{-\frac{1}{2}}[\alpha(1)\beta(2) - \beta(1)\alpha(2)]. \qquad \textbf{1.5}$

Notice that two other symmetric spin functions need to be considered,

$$\psi_{spin}(+)_{(ii)} = \alpha(1)\alpha(2) \qquad \textbf{1.6}$$

and $\quad \psi_{spin}(+)_{(iii)} = \beta(1)\beta(2). \qquad \textbf{1.7}$

1.4 Energy states in the helium atom

The Hamiltonian operator for the helium atom will contain the following terms (Z, the atomic number, will be 2 for helium):

$$\left\{ -\frac{h^2}{8\pi^2 m}\nabla^2(1) - \frac{Ze^2}{r(1)} - \frac{h^2}{8\pi^2 m}\nabla^2(2) - \frac{Ze^2}{r(2)} + \frac{e^2}{r(12)} \right\},$$

which may be written more compactly as,

$$\left\{ \mathscr{H}(1) + \mathscr{H}(2) + \frac{e^2}{r(12)} \right\}.$$

The first two terms comprise those parts of the Hamiltonian referring to electrons 1 and 2 exclusively. The third term involves repulsion between electrons 1 and 2. Let us now evaluate the ground state energy.

$$E = \frac{\frac{1}{2}\iint 1s(1)1s(2)[\alpha(1)\beta(2) - \beta(1)\alpha(2)][\mathscr{H}(1) + \mathscr{H}(2) + e^2/r(12)]1s(1)1s(2)}{\frac{1}{2}\iint 1s(1)1s(2)[\alpha(1)\beta(2) - \beta(1)\alpha(2)]1s(1)1s(2)[\alpha(1)\beta(2) - \beta(1)\alpha(2)] \, d\tau(1) \, d\tau(2)} \times$$

$$\times [\alpha(1)\beta(2) - \beta(1)\alpha(2)] \, d\tau(1) \, d\tau(2). \qquad \textbf{1.8}$$

The terms involving $\mathscr{H}(1)$ on the top line are:

$\int 1s\beta(2)1s\beta(2) \, d\tau(2) \int 1s\alpha(1)[\mathscr{H}(1)]1s\alpha(1) \, d\tau(1) -$
$$\text{A}$$

$-\int 1s\beta(2)1s\alpha(2) \, d\tau(2) \int 1s\alpha(1)[\mathscr{H}(1)]1s\beta(1) \, d\tau(1) -$
$$\text{B}$$

$-\int 1s\alpha(2)1s\beta(2) \, d\tau(2) \int 1s\beta(1)[\mathscr{H}(1)]1s\alpha(1) \, d\tau(1) +$
$$\text{C}$$

$+\int 1s\alpha(2)1s\alpha(2) \, d\tau(2) \int 1s\beta(1)[\mathscr{H}(1)]1s\beta(1) \, d\tau(1).$
$$\text{D}$$

The four terms are written $\text{A} - \text{B} - \text{C} + \text{D}$. In all cases functions involving the same electron are collected together for integration. The first part of integral A concerns electron 2. If we assume that the spin functions represented by α and β are both normalized and orthogonal then the spin part of this integral is $+1$; so also is the orbital part. The whole of the first part of A is therefore $+1$. In the second part the spin function, this time for electron 1, will also be $+1$. Thus A reduces to $\int 1s(1)[\mathscr{H}(1)]1s(1) \, d\tau(1)$ which is simply the energy of an electron in the $1s$ orbital all by itself. In term **B** the spin functions for electron 2 are different, since α and β represent orthogonal functions the whole term will be zero; **C** will also be zero for the same reason. **D** is a similar term to **A** and reduces again to the energy of an electron in a $1s$ orbital.

Returning now to the original equation **1.8** and considering the action of $\mathscr{H}(2)$, this will act in much the same way as $\mathscr{H}(1)$ and again yield two terms each representing the energy of an electron in a $1s$ orbital. The terms containing the electron repulsion factor are

$$+ \iint 1s\alpha(1)1s\beta(2)\left[\frac{e^2}{r(12)}\right]1s\alpha(1)1s\beta(2)\,d\tau(1)\,d\tau(2)-$$
$$\underset{\mathsf{E}}{}$$

$$- \iint 1s\alpha(1)1s\beta(2)\left[\frac{e^2}{r(12)}\right]1s\beta(1)1s\alpha(2)\,d\tau(1)\,d\tau(2)-$$
$$\underset{\mathsf{F}}{}$$

$$- \iint 1s\beta(1)1s\alpha(2)\left[\frac{e^2}{r(12)}\right]1s\alpha(1)1s\beta(2)\,d\tau(1)\,d\tau(2)+$$
$$\underset{\mathsf{G}}{}$$

$$+ \iint 1s\beta(1)1s\alpha(2)\left[\frac{e^2}{r(12)}\right]1s\beta(1)1s\alpha(2)\,d\tau(1)\,d\tau(2).$$
$$\underset{\mathsf{H}}{}$$

Of these terms **F** and **G** will be zero because of spin orthogonality. The electron repulsion operator will interact with orbital functions which describe electron distribution, but not with spin terms. Normalized spin functions may be removed from **E** and **H**, which then become

$$\mathsf{E} = \mathsf{H} = \iint 1s(1)1s(2)\left[\frac{e^2}{r(12)}\right]1s(1)1s(2)\,d\tau(1)\,d\tau(2).$$

Such integrals are given the symbol J and called *Coulomb integrals* because they measure the Coulombic repulsion force between electrons, in this case, an electron in orbital $1s$ and another electron in the same orbital (see problem 1.3).

Thus $E = \frac{1}{2}[2E_{1s}(1)+2E_{1s}(2)+2J]$

$$= E_{1s}(1)+E_{1s}(2)+J.$$

($E_{1s}(i)$ = energy of the ith electron in the $1s$ orbital of helium.)

This result is entirely reasonable. Using the ideas of spin or orbital orthogonality that were so useful in simplifying the terms in the equations for the ground state let us now investigate the first excited state. The following wave functions are possible (referring to the equations given on p. 20):

$\Psi_{\mathrm{I}} = \psi_{\mathrm{orb}}(+)\times\psi_{\mathrm{spin}}(-),$ **(1.2 and 1.5)**

$\Psi_{\mathrm{II}} = \psi_{\mathrm{orb}}(-)\times\psi_{\mathrm{spin}}(+)_{(\mathrm{i})},$ **(1.3 and 1.4)**

$\Psi_{\mathrm{III}} = \psi_{\mathrm{orb}}(-)\times\psi_{\mathrm{spin}}(+)_{(\mathrm{ii})},$ **(1.3 and 1.6)**

$\Psi_{\mathrm{IV}} = \psi_{\mathrm{orb}}(-)\times\psi_{\mathrm{spin}}(+)_{(\mathrm{iii})}.$ **(1.3 and 1.7)**

Thus $\Psi_I = \frac{1}{2}[1s\alpha(1)2s\beta(2) - 1s\beta(1)2s\alpha(2) + 2s\alpha(1)1s\beta(2) - 2s\beta(1)1s\alpha(2)]$,

$$\qquad\qquad \text{L} \quad - \quad \text{M} \quad + \quad \text{N} \quad - \quad \text{P}$$

$\Psi_{II} = \frac{1}{2}(L + M - N - P)$,

$\Psi_{III} = 2^{-\frac{1}{2}}[1s\alpha(1)2s\alpha(2) - 2s\alpha(1)1s\alpha(2)]$,

$\Psi_{IV} = 2^{-\frac{1}{2}}[1s\beta(1)2s\beta(2) - 2s\beta(1)1s\beta(2)]$

(see problem 1.4).

If we now evaluate $\int \Psi_I \mathcal{H} \Psi_I \, d\tau$ we shall also be able to find $\int \Psi_{II} \mathcal{H} \Psi_{II} \, d\tau$ by changing the signs of the terms M and N.

Assuming that Ψ_1 is normalized,

$$E_I = \frac{1}{4} \iint (L - M + N - P) \left[\mathcal{H}(1) + \mathcal{H}(2) + \frac{e^2}{r(12)} \right] (L - M + N - P) \, d\tau(1) \, d\tau(2).$$

The operator may be broken down and its three parts considered in turn.

For $\mathcal{H}(1)$ $\quad \iint [L\mathcal{H}(1)L + M\,\mathcal{H}(1)M + N\,\mathcal{H}(1)N + P\,\mathcal{H}(1)P] \, d\tau(1) \, d\tau(2).$

All possible cross terms, for example $\iint L\mathcal{H}(1)N \, d\tau(1) \, d\tau(2)$, reduce to zero because of either spin or orbital orthogonality of electron 2. Using the same symbolism as before the four terms can be written $2E_{1s}(1) + 2E_{2s}(1)$. Similarly for the operator of $\mathcal{H}(2)$ the result is $2E_{1s}(2) + 2E_{2s}(2)$.

When the electron repulsion part of the operator is considered only spin orthogonality can be used to help reduce the number of terms since all orbitals for both electrons will be connected with the operation of $e^2/r(12)$. The four terms L( )L, M( )M, N( )N and P( )P will obviously occur where ( ) symbolizes $e^2/r(12)$. Note also that the following terms will *not* be zero since the spin functions for (1) and (2) are the same in both parts of the wave function, $+$L( )N, $+$M( )P, $+$N( )L and $+$P( )M.

Let us now examine typical integrals in more detail.

$$\iint L( )L \, d\tau(1) \, d\tau(2) = \iint 1s\alpha(1)2s\beta(2) \left[\frac{e^2}{r(12)} \right] 1s\alpha(1)2s\beta(2) \, d\tau(1) \, d\tau(2).$$

When the spin functions have been removed this is clearly an integral of the type J, here describing the repulsion between an electron 1 in a 1s orbital and an electron 2 in a 2s orbital.

Integrals of the type L( )N are

$$\iint L( )N \, d\tau(1) \, d\tau(2) = \iint 1s\alpha(1)2s\beta(2) \left[\frac{e^2}{r(12)} \right] 2s\alpha(1)2s\beta(2) \, d\tau(1) \, d\tau(2).$$

Again the spin terms can be removed and the integral reduces to

$$\iint L( )N \, d\tau(1) \, d\tau(2) = \iint 1s(1)2s(2) \left[\frac{e^2}{r(12)} \right] 2s(1)1s(2) \, d\tau(1) \, d\tau(2).$$

This again involves repulsion between two electrons but electron 1 is associated with both functions 1s and 2s, and so is electron 2. Because it is as though the electrons were being exchanged between these two orbitals, integrals of this type are called *exchange integrals*; they are given the symbol K.

The total energy associated with Ψ_I is therefore

$$E_I = \tfrac{1}{4}[2E_{1s}(1) + 2E_{2s}(1) + 2E_{1s}(2) + 2E_{2s}(2) + 4J_{1s2s} + 4K_{1s2s}]$$

$$= E_{1s} + E_{2s} + J_{1s2s} + K_{1s2s}.$$

Both J and K are electron repulsion terms and so will tend to make the energy less than that estimated merely by inspection of the configuration $1s^1 2s^1$.

The reader should now follow the working through Ψ_{II} (see problem 1.5); the only difference will be that M and N have changed signs. The effect is only seen in the K integral; the energy for Ψ_{II} is

$$E_{II} = E_{1s} + E_{2s} + J_{1s2s} - K_{1s2s}.$$

The same energy will also be found for Ψ_{III} and Ψ_{IV} (see problem 1.5). These were all associated with the $\psi_{spin}(+)$ function. When two or more wave functions, such as Ψ_{II}, Ψ_{III} and Ψ_{IV} in this case, have the same energy they are said to be *degenerate*. This type of degeneracy, associated with spin functions is called *multiplicity*. Different multiplicities have different names, for a multiplicity of one, that is only one spin function – singlet; multiplicity two (two spin functions) – doublet; multiplicity three (as for Ψ_{II}, Ψ_{III} and Ψ_{IV}) – triplet; four – quartet; five – quintet; etc.

Thus although four possible wave functions have been found for the excited configuration of helium $1s^1 2s^1$ there are only two distinct energies, one associated with the singlet wave function Ψ_I and the other associated with the trio of wave functions Ψ_{II}, Ψ_{III} and Ψ_{IV} that make up the triplet. The energy difference between the singlet and triplet states is $2K_{1s2s}$ and since both J and K represent electron repulsion terms the triplet state is the more stable. This result can be generalized, that states of the highest multiplicity will be the most stable (Hund's rule).

The form of the arguments presented above is quite general and shows how electron repulsion terms enter into the calculations of energy in both atomic and molecular systems. They also demonstrate the importance of electron spin and the effect which it too can have upon the energy. The highest multiplicity in a system will be associated with the maximum number of electrons of the same spin. Since such electrons cannot be distinguished by m_s they must be distinguished by m_l (assuming n and l to be the same). Thus electrons of the same spin must be in different orbitals and must stay in different orbitals. This has the effect of keeping such electrons apart and reducing electron repulsion between them. In the helium example an electron repulsion term K_{1s2s} was added to the total energy reducing the stability of the system; but in the corresponding triplet state the stability was enhanced by the same amount.

It is usually found that integrals of the type J are much larger than those of type K. Also the values of J-integrals vary for repulsions between orbitals of different n and l values. This is particularly important in polyelectronic atoms since J_{1s2s} and J_{1s2p}, for example, will have different values. Thus the degeneracy associated with different values of l for the same value of n found in the one-electron hydrogen atom is broken down and quite large energy differences can be found between s, p, d, etc. orbitals which have the same principal quantum number. With increasing atomic number electrons may be thought of as being added to the atom: fed into orbitals which are basically hydrogen like. Thus the first two electrons will go to the $1s$ orbital, then to the $2s$ which, being no longer of the same energy with $2p$ orbitals, will be filled first. Boron will therefore have a ground state configuration $1s^2 2s^2 2p^1$. Degeneracy of orbitals of the same n and l but differing only in m_l is destroyed only by a magnetic field so the p-orbitals are degenerate and a variety of spin and orbital function combinations is possible for configurations such as $1s^2 2s^2 2p^2$ (see Chapter 6). The energies of the states associated with the configuration $1s^2 2s^1 2p^3$ are distinctly higher (see Figure 5), demonstrating

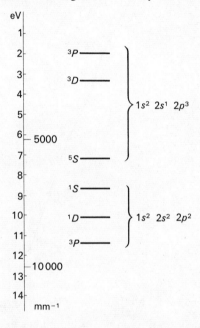

Figure 5   Spectroscopic states of carbon derived from various configurations

that in this polyelectronic system $2p$ orbitals are less tightly bound than $2s$ orbitals. This ns–np energy difference increases rapidly on going to the right in the periodic table (Figure 6). It is important to remember the existence of this energy difference when considering chemical bonding since one of the criteria for strong covalent bonding is that the participating orbitals shall have comparable energy.

1.5   The correlation of electrons

So far in our discussion of polyelectronic systems, and helium in particular, we have considered in some detail the nature of the Hamiltonian operator (e.g. the inclusion of electron repulsion terms) and the form of the wave function (i.e.

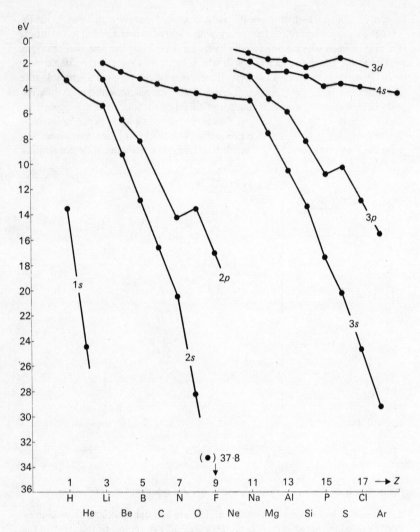

Figure 6   Energy levels of various orbitals for the first eighteen elements

Slater determinant). This had led us to the conclusion that a particular electronic configuration may give rise to a variety of states of different energies. This differentiation is due to the inclusion of spin terms in the wave function. Thus in the $1s^1 2s^1$ configuration of helium the symmetric spin function is associated with a more stable state than the antisymmetric spin function. In the former state the electrons are constrained to keep apart more than in the latter. The general tendency of electrons to avoid each other is called *electron correlation*. It is always simply a manifestation of the mutual repulsion of the charges. However, when

26   Atoms

this effect is brought into play as a result of the explicit consideration of electron spin, as above, it is referred to as *spin correlation*.

It is now important to at least consider the effect of interelectron repulsions upon the orbitals in the determinant wave function. Surely it is as logical to include this effect as it is to include the electron repulsion term in the Hamiltonian operator? Qualitatively it is fairly easy to see what will happen. Consider two electrons in the $1s$ orbital of the helium atom. The simplest orbital for each electron would be a $1s$ hydrogen-like orbital, adapted for the increased nuclear charge of the helium nucleus. However, the two electrons will also tend to repel each other and so will, more often than not, be found on opposite sides of the nucleus. This effect can be called *charge correlation*. The 'correlated' pair of electrons will, of course, move freely in the space around the nucleus so that, over a period of time, a spherically symmetric distribution of charge would be found, as required of a $1s$ wave function. The point is that the probability of the electrons coming close together is less than would be expected by the super-position of two independent $1s$ wave functions. Since it is functions of this type that have been used implicitly in the calculations for the helium atom the electron repulsion integrals J will be rather too large.

The next step is to develop wave functions corrected for electron correlation. This is a matter of some difficulty and complexity so only two possible approaches will be outlined here. In the first the Slater determinant (using hydrogen-like orbitals) is multiplied by a factor which would become small when the inter-electron distance is small. As a naïve example for the ground state of helium a correlation factor could be something like $r(12)$ or $\exp r(12)$. Another approach is to retain the Slater determinant using hydrogen-like wave functions but to con-sider possible interactions between electronically excited states of the right sym-metry – *configuration interaction*. Usually the charge distribution in such states is quite different so that this sort of interaction results in a redistribution of charge. For example one of the excited states of helium with two electrons of opposite spin in the same p-orbital would be able to interact with the ground state. Since p-orbitals have axial rather than spherical symmetry the interaction between these two states will introduce some p (axial) character into the simple $1s^2$ description of the ground state. This is the same as enhancing the probability that electrons will be found on opposite sides of the nucleus.

Electron correlation is particularly important in understanding the shapes of molecules.

No new principles will be met in considering atoms more complicated than helium, or even molecules, but the complexity of the calculations increases enor-mously. We shall leave such discussions to more specialized texts and turn now to the formation of a simple covalent chemical bond.

Problems 1

One electron volt is equivalent to $806 \cdot 8$ mm^{-1}, *to* $23 \cdot 1$ kcal mol^{-1} *and to* $1 \cdot 60 \times 10^{-19}$ J *per molecule (or* $96 \cdot 5$ kJ mol^{-1}).

1.1 The Schrödinger equation in one dimension (zero potential energy) is

$$\left[-\frac{h^2}{8\pi^2 m} \frac{\partial^2}{\partial x^2} \right] \psi = E\psi.$$

If a particle (mass m) is constrained to move along a line of length a metres, show that $\psi = A \sin Bx$ is an eigenfunction. Determine the constants A and B and find the possible eigenvalues.

B can be found using the 'boundary conditions', that is ψ must be 0 at $x = 0$ and at $x = a$, since the particle only moves within these limits. A can be found by requiring that ψ be normalized.

If the particle has the mass of an electron ($\sim 9 \times 10^{-31}$ kg) and a has the length of the π molecular orbital system in buta-1,3-diene, $0 \cdot 42$ nm ($= 0 \cdot 42 \times 10^{-9}$ m), find the three lowest eigenvalues (in electron volts). Thus determine the energy difference between the second and the third eigenvalue. Compare this value with the energy of the $\pi \rightarrow \pi^*$ transition of buta-1,3-diene at about 6 eV.

1.2 Consider a particle to move on the circumference of a circle of radius r. The distance along an arc is $r\phi$ which subtends ϕ at the centre. The motion of the particle can therefore be described in terms of this one dimension. Deduce the appropriate form of the Schrödinger equation ($V = 0$) using $x = r\phi$.

Show that $\quad \psi = C \sin D\phi$

and $\quad\quad\quad \psi' = C' \cos D'\phi$

are both acceptable wave functions and determine C, D, C' and D'. The 'boundary conditions' will be that after rotation through 2π, ψ or ψ' should repeat itself. Show that all possible eigenfunctions of ψ and ψ' are orthogonal to each other.

If the particle has electronic mass and the circle has the radius of the benzene molecule ($0 \cdot 14$ nm) find the energies of the first three ψ' functions and the first two ψ functions.

1.3 Show that the wave function used for the helium atom is normalized, that is that the lower line of equation **1.8** is equal to $+1$.

1.4 Express Ψ_I, Ψ_{II}, Ψ_{III} and Ψ_{IV} wavefunctions for the $1s^1 2s^1$ excited state of helium as Slater determinants.

1.5 Evaluate the energy of the function Ψ_{II} (p. 23) and verify the result given on p. 24. Show that Ψ_{III} and Ψ_{IV} both have the same energy as Ψ_{II}.

Chapter 2
Molecules

2.1   The chemical bond

If two atoms approach each other the electrons of each atom will experience increased attraction due to the nucleus of the other. If this increased attraction outweighs the increase in interelectron repulsion which will also occur and if suitable orbitals are available then a new system, more stable than the original free atoms (i.e. ΔG is negative for the formation of the molecule), is formed in which the two atoms are linked by a chemical bond. Clearly the energy of the electrons in the new environment, especially when compared with the old, will be of great importance in determining the properties of the bond. In order to calculate this energy we need to make some assumptions about suitable wave functions and the type of Hamiltonian operator to be used.

2.1.1   *Wave functions*

An over-all wave function of the determinant type would clearly be acceptable but the question arises, what sort of wave function can be used to describe the orbital motion of each electron in the whole molecule? A simple, approximate and very useful answer is to regard each molecular orbital as the sum of various amounts of atomic orbitals thus,

$$\psi_i = \sum_r a_{ri} \phi_r,$$

where ψ_i is the *i*th *molecular* orbital, ϕ_r is the *r*th *atomic* orbital and a_{ri} is a coefficient describing the amount of atomic orbital *r* to be found in the molecular orbital *i*.

Molecular orbitals composed in this way are said to be linear combinations of atomic orbitals (L.C.A.O.).

No account is taken of the possible interactions (deformation, polarization etc.) which may exist between constituent orbitals and to this extent the L.C.A.O. method is approximate; even so, because of its simplicity it is most useful.

Each molecular orbital of this type can be associated with a particular spin and used in the determinantal wave functions just like an atomic orbital. It is now of some importance to try and determine the values of the coefficients a_{ri} in the above equations. These must correspond to stable orbitals in molecules. A simple way to evaluate these coefficients would therefore be to study the variation of the energy E_i of a molecular orbital with variation of each of the coefficients

a_{ri} in turn, and noting the values of a_{ri} which gave a minimum value of E_i. Such a study is easily made by partial differentiation and equating the various differentials to zero, that is $\partial E_i / \partial a_{ri} = 0$ for a minimum energy. This technique is known as the variation method. Before we can carry out this work we must look at the Hamiltonian operator we shall use to find the energies of these molecular orbitals.

2.1.2 *Hamiltonian operator*

For a system of two nuclei, A and B, and two electrons the operator will be similar to that for helium:

$$\left[-\frac{h^2}{8\pi^2 m}\nabla^2(1) - \frac{Z_A e^2}{r(1A)} - \frac{Z_B e^2}{r(1B)} - \frac{h^2}{8\pi^2 m}\nabla^2(2) - \frac{Z_A e^2}{r(2A)} - \frac{Z_B e^2}{r(2B)} + \frac{e^2}{r(12)} + \frac{Z_A Z_B e^2}{r(AB)} \right]$$

The new terms that have been introduced describe attraction of both electrons to both nuclei and also the electrostatic repulsion between the two nuclei.

For a system of n electrons and x nuclei the operator may be generalized,

$$\left[\sum_{i=1}^{n} \frac{h^2}{8\pi^2 m}\nabla^2(i) - \sum_{C=1}^{x} \sum_{i=1}^{n} \frac{Z_C e^2}{r(iC)} + \sum_{i>j} \sum \frac{e^2}{r(ij)} + \sum_{C>D} \sum \frac{Z_C Z_D e^2}{r(CD)} \right],$$

where i, j are electrons, C, D are nuclei, $r(iC)$ is the distance between electron i and nucleus C, $r(ij)$ is the distance between electron i and electron j and $r(CD)$ is the distance between nucleus C and nucleus D.

Note that the summation of the repulsion terms is arranged so that each repulsion is counted once only and not twice. This operator can be written in a more simple form,

$$\left[\sum_{i=1}^{n} \mathscr{H}(i) + \sum_{i>j} \sum \frac{e^2}{r(ij)} + P \right],$$

where the symbol $\mathscr{H}(i)$ collects together all the operations that act upon the ith electron alone. The second term is the electron repulsion term between the ith and jth electrons and P represents repulsions between nuclei. For a given arrangement of nuclei P is assumed constant.

Clearly the evaluation of the energies of molecular orbitals using this operator and determinantal wave functions will be a matter of some complexity. Let us therefore seek approximations that will simplify the task. The most awkward term in the Hamiltonian operator is the electron repulsion term which links every possible pair of electrons. This term can either be completely ignored or else each $\mathscr{H}(i)$ adapted in such a way as to try and take account of electron repulsion in an averaged-out sort of way. Let us symbolize such repulsion-adapted one-electron operators as $\mathscr{H}'(i)$ and also let us ignore P in the subsequent discussions since it will not affect electronic energies but only the total energy of the system. The new operator is therefore

$$\left[+ \sum_{i=1}^{n} \mathscr{H}'(i) \right]. \qquad \textbf{2.1}$$

The calculations on p. 21 with various electronic configurations of the helium atom showed that the total energy of the system was simply the sum of the orbital energies, provided only that the one-electron operators (e.g. $\mathscr{H}(1)$, $\mathscr{H}(2)$) were used. More complex J and K terms arose only from explicit consideration of electron repulsion between two electrons.

The total energy of a molecule can therefore, at this level of approximation, be broken down into the sum of the energies of the occupied molecular orbitals. The problem now is considerably simplified, to that of finding the energies of the individual molecular orbitals. The energy can be found in the usual way,

$$E = \frac{\int \psi \mathscr{H} \psi^* \, d\tau}{\int \psi \psi^* \, d\tau}.$$

$\mathscr{H}$ in this equation will be $\mathscr{H}'$ for a typical electron as defined above (2.1) and ψ the wave function for any molecular orbital. If we now express ψ in terms of atomic orbitals using the L.C.A.O. approximation the equation becomes

$$
\begin{aligned}
E &= \frac{\int (\sum_r a_r \phi_r) \, \mathscr{H}' (\sum_s a_s \phi_s) \, d\tau}{\int (\sum_r a_r \phi_r)(\sum_s a_s \phi_s) \, d\tau} \\
&= \frac{\sum_r \sum_s a_r a_s \int \phi_r \mathscr{H}' \phi_s \, d\tau}{\sum_r \sum_s a_r a_s \int \phi_r \phi_s \, d\tau}.
\end{aligned}
$$

Different subscripts r and s have been used to ensure the complete expansion of ψ wherever it occurs.

Now quite obviously E will be able to take on a wide variety of values depending upon values of a_r and a_s but we shall only be interested in those values of E that correspond to a stable molecular orbital, that is when $(\partial E/\partial a_t) = 0$ as discussed above; where a_t is a typical coefficient.

Let the equations for E be written

$$E = \frac{u}{v},$$

then

$$\frac{\partial E}{\partial a_t} = \frac{v \, \partial u/\partial a_t - u \, \partial v/\partial a_t}{v^2}.$$

For a minimum value of E,

$$\frac{\partial E}{\partial a_t} = 0,$$

thus

$$\left(\frac{u}{v}\right)_{\min} = \frac{\partial u/\partial a_t}{\partial v/\partial a_t}.$$

But u/v is the energy E,

hence $\quad E_{min} = \dfrac{\partial u/\partial a_t}{\partial v/\partial a_t}.$

To evaluate $\partial u/\partial a_t$:

$$u = \sum_r \sum_s a_r a_s \int \phi_r \mathscr{H}' \phi_s \, d\tau,$$

$$u = \sum_{r \neq t} a_r a_t \int \phi_r \mathscr{H}' \phi_t \, d\tau + \sum_{s \neq t} a_t a_s \int \phi_t \mathscr{H}' \phi_s \, d\tau + \sum_{r \neq t} \sum_{s \neq t} a_r a_s \int \phi_r \mathscr{H}' \phi_s \, d\tau + (a_t^2) \int \phi_t \mathscr{H}' \phi_t \, d\tau,$$

$$\frac{\partial u}{\partial a_t} = \sum_{r \neq t} a_r \int \phi_r \mathscr{H}' \phi_r \, d\tau + \sum_{s \neq t} a_s \int \phi_t \mathscr{H}' \phi_s \, d\tau + 2a_t \int \phi_t \mathscr{H}' \phi_t \, d\tau.$$

In this expression there is no longer any need to distinguish between subscripts r and s; r alone will therefore be used. Clearly $\partial v/\partial a_t$ can be evaluated in an exactly similar way, so that

$$E_{min} = \frac{2 \sum_r a_r \int \phi_r \mathscr{H}' \phi_t \, d\tau}{2 \sum_r a_r \int \phi_r \phi_t \, d\tau}.$$

Let us now drop the subscript 'min' and write the integrals in the following way:

$$\int \phi_r \mathscr{H}' \phi_t \, d\tau = H_{rt},$$

$$\int \phi_r \phi_t \, d\tau = S_{rt}.$$

Then $\quad\quad\quad\quad E = \dfrac{\sum\limits_r a_r H_{rt}}{\sum\limits_r a_r S_{rt}}$

and $\quad \sum\limits_r a_r (H_{rt} - E S_{rt}) = 0.$ **2.2**

If there are n atomic orbitals then both r and t will run to n terms. Equation **2.2** is therefore a concise expression for the following array of simultaneous equations:

$t = 1 \quad a_1(H_{11} - ES_{11}) + a_2(H_{21} - ES_{21}) + \ldots + a_n(H_{n1} - ES_{n1}) = 0,$

$t = 2 \quad a_1(H_{12} - ES_{12}) + a_2(H_{22} - ES_{22}) + \ldots + a_n(H_{n2} - ES_{n2}) = 0,$

$\vdots \quad\quad\quad \vdots \quad\quad \vdots \quad\quad\quad \vdots \quad\quad\quad\quad \vdots \quad\quad\quad \vdots$

$t = n \quad a_1(H_{1n} - ES_{1n}) + a_2(H_{2n} - ES_{2n}) + \ldots + a_n(H_{nn} - ES_{nn}) = 0.$

These equations are known as *secular equations*. Such an array of linear homogeneous simultaneous equations will only yield non-trivial values for the a_is (i.e. all a_is equal to zero) when the determinant of the coefficients itself is zero. The solution of this *secular determinant* will yield particular values of E in terms of the parameters H_{rt} and S_{rt}. These will be the eigenvalues and correspond to the energies of the molecular orbitals. These eigenvalues may then be inserted in the secular equations to yield the values of the coefficients a_i. The secular determinant corresponding to the above array of secular equations is

$$\begin{vmatrix} H_{11}-ES_{11} & H_{21}-ES_{21} & \cdots & H_{n1}-ES_{n1} \\ H_{12}-ES_{12} & H_{22}-ES_{22} & \cdots & H_{n2}-ES_{n2} \\ \vdots & \vdots & & \vdots \\ H_{1n}-ES_{1n} & H_{2n}-ES_{2n} & \cdots & H_{nn}-ES_{nn} \end{vmatrix} = 0.$$

To proceed further it is necessary to consider the actual values to be used for H_{rt} and S_{rt}. The following terms and symbols are commonly used:

$H_{rr} = \alpha_r = $ Coulomb integral (not to be confused with J, the electron repulsion Coulomb integral),

$H_{rt} = \beta_{rt} = $ resonance integral,

$S_{rr} = +1$ if ϕ_r normalized,

$S_{rt} = $ overlap integral.

Further approximations can now be introduced that make the solution of the secular determinant relatively easy. The first – due to Hückel – is to neglect resonance integrals between orbitals on non-adjacent atoms. The second – the 'neglect of overlap' approximation – involves, as its name suggests, putting all overlap integrals equal to zero (i.e. $S_{rt} = 0$, $r \neq t$). This second approximation sounds pretty drastic, and indeed downright illogical, for if orbitals do not overlap, how can there be a non-zero resonance integral? Even so the neglect of overlap works surprisingly well and does not seem to introduce greater errors into the calculations than those that have gone before. Let us now look at the evaluation of some molecular orbital energies using this method.

2.2 Examples of simple molecular orbital calculations

2.2.1 π-Orbitals in the allyl radical

Consider only the three p-orbitals as shown in Figure 7. Using the approximations suggested above the secular determinant is

$$\begin{vmatrix} \alpha_1-E & \beta_{21} & 0 \\ \beta_{12} & \alpha_2-E & \beta_{32} \\ 0 & \beta_{23} & \alpha_3-E \end{vmatrix} = 0.$$

$\alpha_1, \alpha_2, \alpha_3$ all refer to the integral $\int \phi_r \mathscr{H}' \phi_r \, d\tau$ where ϕ_p is a $2p$ orbital on a carbon atom. Clearly the αs will not be the same as the ionization potential of an electron in a $2p$ orbital since the Hamiltonian $\mathscr{H}'$ contains some terms that attempt to deal with electron repulsion in the molecule. Such repulsions might well be different at the central atom and the terminal atoms. Nevertheless let us ignore these possible differences and set $\alpha_1 = \alpha_2 = \alpha_3 = \alpha$.

It is perhaps more reasonable to suggest that all the non-zero resonance integrals should be identical if the C_1-C_2 and C_2-C_3 bond lengths are the

33   Examples of Simple Molecular Orbital Calculations

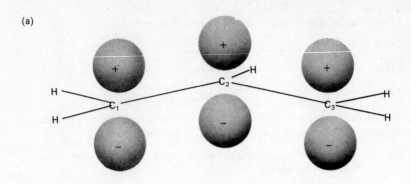

(a)

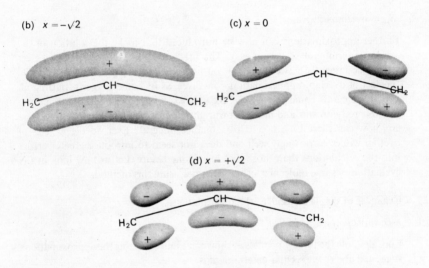

(b) $x = -\sqrt{2}$

(c) $x = 0$

(d) $x = +\sqrt{2}$

Figure 7   Formation of π-orbitals in the allyl radical from three carbon $2p$ atomic orbitals: (a) carbon p-orbitals; (b) bonding, (c) nonbonding, (d) antibonding π molecular orbitals

same. The determinant can be simplified

$$\begin{vmatrix} \alpha - E & \beta & 0 \\ \beta & \alpha - E & \beta \\ 0 & \beta & \alpha - E \end{vmatrix} = 0,$$

or, putting $(\alpha - E)/\beta = x$,

$$\begin{vmatrix} x & 1 & 0 \\ 1 & x & 1 \\ 0 & 1 & x \end{vmatrix} = 0,$$

i.e. $x(x^2-1)-(x) = 0 = x(x^2-2)$,

$x = 0, \pm\sqrt{2}$.

Now $E = \alpha - x\beta$,

i.e. $E = \alpha - (-\sqrt{2})\beta = \alpha + \sqrt{2}\beta$ (bonding),

$E = \alpha - (0)\beta = \alpha$ (nonbonding),

or $E = \alpha - (+\sqrt{2})\beta = \alpha - \sqrt{2}\beta$ (antibonding),

$\left.\right\}$ π-molecular orbital energies.

This result shows that the type of interaction between three atomic orbitals considered in the allyl system yields three molecular orbitals. Electrons in the first ($x = -\sqrt{2}$) will be more tightly bound, by $\sqrt{2}\beta$, than in an original atomic orbital. Electrons in the second ($x = 0$) will be bound with the same energy as in an atomic orbital and electrons in the third ($x = -\sqrt{2}$) will be less tightly bound. Hence orbitals of the first type are called *bonding*, the second, *nonbonding* and the third, *antibonding*. It is really quite remarkable that this result has been achieved so simply and without recourse to any quantitative data. It is even possible to make predictions of chemical interest. Since the second orbital is nonbonding the π-energies (of (allyl)$^+$, (allyl)$\cdot$ and (allyl)$^-$), relative to electrons in $2p$ orbitals of free atoms, will be the same. This conclusion should be treated as qualitative rather than quantitative because of the large number of approximations introduced into the calculations.

Now that the energies of the various molecular orbitals have been found let us calculate the atomic orbital coefficients in each one of them. The secular equations for the π allyl orbitals will be (using the approximations and symbols introduced above)

$a_1 x + a_2 = 0,$ **2.3**

$a_1 + a_2 x + a_3 = 0,$ **2.4**

$a_2 + a_3 x = 0.$

To find the values of a_r for the bonding orbital, the value $x = -\sqrt{2}$ is substituted in the above equations.

Substituting in equation **2.3**:

$-a_1\sqrt{2} + a_2 = 0,$

$a_2 = a_1\sqrt{2}.$

Substituting in equation **2.4**:

$a_1 + (a_1\sqrt{2})(-\sqrt{2}) + a_3 = 0,$

$a_3 = a_1.$

Thus all the coefficients are known in terms of a_1.

Now we must use the condition that the molecular orbital shall be normalized,

i.e. $\sum_{r=1}^{n} a_r^2 = +1$.

For allyl $(x = -\sqrt{2})$ $\quad a_1^2 + a_2^2 + a_3^2 = +1$,

i.e. $\qquad\qquad\qquad a_1^2 + 2a_1^2 + a_1^2 = +1$.

Thus $\quad a_1 = \frac{1}{2}$,

$\qquad a_2 = \sqrt{\frac{1}{2}}$,

$\qquad a_3 = \frac{1}{2}$.

Similar calculations can be performed for the other orbitals.

For $x = 0$ $\qquad a_1 0 + a_2 = 0$,
$$a_2 = 0;$$
$$a_1 + a_2 0 + a_3 = 0,$$
$$-a_1 = +a_3.$$

Normalization:

$$a_1^2 + 0^2 + (-a_1)^2 = +1.$$

Hence $\qquad a_1 = +\sqrt{\frac{1}{2}}$

and $\qquad a_3 = -\sqrt{\frac{1}{2}}$.

When $x = +\sqrt{2}$ $\qquad a_1 \sqrt{2} + a_2 = 0$,
$$a_2 = -\sqrt{2} a_1;$$
$$a_1 + (-\sqrt{2} a_1)\sqrt{2} + a_3 = 0,$$
$$a_3 = +a_1.$$

Normalization:

$$a_1^2 + (-\sqrt{2} a_1)^2 + a_1^2 = +1.$$

Hence $\qquad a_1 = +\frac{1}{2}$,

$\qquad a_2 = -\sqrt{\frac{1}{2}}$,

and $\qquad a_3 = +\frac{1}{2}$.

Figure 7 shows diagrammatic representations of these three molecular orbitals.

The probability of finding an electron at a given point in space is $\psi\psi^*$ or, if we ignore the possible complex nature of the wave function, ψ^2. If we now expand the molecular orbital in terms of its atomic constituents, and ignore overlap then,

$$\int \psi_i \psi_i \, d\tau = \int (\sum_r a_{ri} \phi_r)(\sum_s a_{si} \phi_s) \, d\tau$$
$$= \sum_r a_{ri}^2 \int \phi_r \phi_r \, d\tau,$$

since all terms with $r \neq s$ will be zero.

The terms a_{ri}^2 may thus be identified with the probability that an electron in ψ_i will in fact be in atomic orbital r, in other words a_{ri}^2 measures the fractional electronic charge at atom r in orbital i. The total *charge* q_r at any particular atomic orbital site of a molecular orbital system can therefore be found by evaluating the following sum

$$q_r = \sum_i n_i a_{ri}^2,$$

where n_i is the number of electrons in the ith molecular orbital.

In a rather similar way, a term to describe the distribution of charge between atoms, the *bond order* p_{rs}, is defined as

$$p_{rs} = \sum_i n_i a_{ri} a_{si}.$$

By way of example, these terms will now be evaluated for some allyl derivatives.

Table 2

i		a_{1i}^2	$a_{1i}a_{2i}$	a_{2i}^2	$a_{2i}a_{3i}$	a_{3i}^2
1	$x = -\sqrt{2}$	0·25	0·354	0·5	0·354	0·25
2	$x = 0$	0·5	0	0	0	0·5
3	$x = +\sqrt{2}$	0·25	−0·354	0·5	−0·354	0·25

	n_1	n_2	n_3	q_1	p_{12}	q_2	p_{23}	q_3
(allyl)$^+$	2	0	0	0·5	0·707	1·0	0·707	0·5
(allyl)·	2	1	0	1·0	0·707	1·0	0·707	1·0
(allyl)$^-$	2	2	0	1·5	0·707	1·0	0·707	1·5

Later we shall see that bond order may often be related to bond length and also that the terms for both bond order and charge in an atomic orbital will be of great importance in discussing electron repulsion effects.

2.2.2 *π-Orbitals in the cyclopropenyl system*

Let us now consider another example involving the interaction of three carbon $2p$ orbitals, but of different geometry from the allyl system. In cyclopropenyl the orbitals originate from carbon atoms at the apices of an equilateral triangle

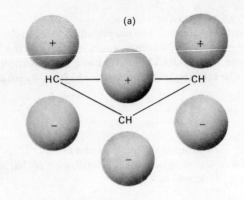

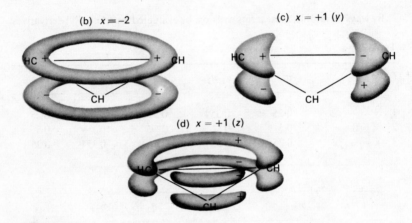

Figure 8   Formation of π-orbitals in the cyclopropenyl system: (a) carbon
p-orbitals; (b) bonding, (c) antibonding, (d) antibonding π molecular orbitals

(Figure 8). Using the same approximations and symbols as for allyl (but not of course implying that the α and β used are necessarily the same for both systems) the secular determinant can be written

$$\begin{vmatrix} x & 1 & 1 \\ 1 & x & 1 \\ 1 & 1 & x \end{vmatrix} = 0.$$

This is equivalent to

$$x(x^2-1)-(x-1)+(1-x) = 0,$$

or $\quad (x-1)(x-1)(x+2) = 0.$

Thus $\qquad\qquad x = -2, +1, +1.$

The corresponding values of E are $\alpha+2\beta$ (bonding), $\alpha-\beta$ (antibonding) and $\alpha-\beta$ (antibonding). Thus a simple difference in molecular geometry has a profound effect upon the nature of the molecular orbitals. In cyclopropenyl just one orbital is much more strongly bonding than the original atomic orbitals and there are two (degenerate) antibonding orbitals. The π-orbital energies of the cyclopropenyl cation, radical and anion will therefore be widely different (4β, 3β and 2β) – an important distinction from the allyl case.

The calculation of the atomic orbital coefficients proceeds as for allyl; thus the original family of secular equations would be

$$a_1 x + a_2 + a_3 = 0,$$

$$a_1 + a_2 x + a_3 = 0,$$

$$a_1 + a_2 + a_3 x = 0.$$

For $x = -2$ $\qquad -2a_1 + a_2 + a_3 = 0$

and $\qquad\qquad\qquad a_1 - 2a_2 + a_3 = 0.$

Thus $\qquad\qquad\qquad -3a_1 + 3a_2 = 0,$

i.e. $\qquad\qquad\qquad\qquad a_1 = a_2.$

Substituting this equality in the first equation,

$$-2a_1 + a_1 + a_3 = 0,$$

$$a_1 = a_3.$$

Normalizing:

$$a_1^2 + a_1^2 + a_1^2 = +1,$$

hence $\quad a_1 = a_2 = a_3 = \sqrt{\tfrac{1}{3}}.$

For $x = +1$ $\qquad a_1 + a_2 + a_3 = 0.$

This same equation results when $x = +1$ is inserted in any one of the above secular equations. Further two molecular orbitals have the same energy, $\alpha-\beta$. A simple, if rather arbitrary, way to calculate the coefficients in a degenerate pair of atomic orbitals such as this is as follows. If a_{rY}, a_{rZ} are the coefficients at atom r in the degenerate pair of molecular orbitals Y and Z and there are n atomic orbitals in all then,

$$a_{rY}^2 + a_{rZ}^2 = \frac{2}{n}, \qquad\qquad\qquad\qquad \textbf{2.5}$$

that is, considered as a pair there is an even charge distribution in the orbitals Y and Z (each normalized to one, hence 2 in the equation above for the pair).

39   Examples of Simple Molecular Orbital Calculations

Further assume that one coefficient in one orbital is zero.

Thus for cyclopropenyl,

if $\quad a_{3Y} = 0,$

then $\quad a_{1Y} + a_{2Y} = 0,$

i.e. $\quad a_{1Y} = -a_{2Y}.$

Normalizing:

$a_{1Y}^2 + (-a_{1Y})^2 + 0^2 = 1$

hence $\quad a_{1Y} = \sqrt{\tfrac{1}{2}},$

$a_{2Y} - \sqrt{\tfrac{1}{2}}.$

If $\quad a_{1Y}^2 + a_{1Z}^2 = \dfrac{2}{n},$

then $\quad \tfrac{1}{2} + a_{1Z}^2 = \tfrac{2}{3},$

i.e. $\quad a_{1Z} = \pm \dfrac{1}{\sqrt{6}}.$

Similarly $\quad a_{2Z} = \pm \dfrac{1}{\sqrt{6}}$

and $\quad a_{3Z} = \pm \sqrt{\dfrac{2}{3}} = \pm \dfrac{2}{\sqrt{6}}.$

Remembering the original secular equation for $x = +1$, $a_1 + a_2 + a_3 = 0$, we see that acceptable coefficients for the Z-orbital are

$$a_{1Z} = +\frac{1}{\sqrt{6}}, \quad a_{2Z} = +\frac{1}{\sqrt{6}}, \quad a_{3Z} = -\frac{2}{\sqrt{6}}.$$

The corresponding coefficients in Y are $a_{1Y} = +\sqrt{\tfrac{1}{2}}, a_{2Y} = -\sqrt{\tfrac{1}{2}}, a_{3Y} = 0$. These orbitals are shown in Figure 8. Note that each contains a nodal plane perpendicular to the plane of the molecule and that these two nodal planes are mutually at right angles. (Degenerate orbitals are discussed in more detail on p. 132.) An interesting property of degenerate orbitals is that linear combinations of these orbitals provide equally good representations. Thus ψ_1' and ψ_1'' may be constructed from ψ_{1Y} and ψ_{1Z} in the following way:

$a\psi_{1Y} + \sqrt{(1-a^2)}\psi_{1Z} = \psi_1',$

$\sqrt{(1-a^2)}\psi_{1Y} - a\psi_{1Z} = \psi_1''.$

Having chosen a as an arbitrary constant $\sqrt{(1-a^2)}$ must be the other coefficient for ψ_1' to be normalized, similarly if $a\psi_{1Y}$ is to be found in ψ_1' then the contribution to ψ_1'' must be $\sqrt{(1-a^2)}$. Convenient forms for these coefficients are sine and cosine functions:

$$(\cos \theta)\psi_{1Y} + (\sin \theta)\psi_{1Z} = \psi_1',$$

$$(\sin \theta)\psi_{1Y} - (\cos \theta)\psi_{1Z} = \psi_1''.$$

The variation of θ is equivalent to a rotation of the pair of nodal planes with respect to the positions of the original atoms. Any possible value of θ will provide an equally valid description of the degenerate orbitals (see problem 2.1).

2.2.3 π-Orbitals in buta-1,3-diene

As a further example of simple molecular orbital calculations let us consider the π-orbitals that would be generated from the overlap of four carbon $2p$ orbitals perpendicular to the plane of the buta-1,3-diene molecule. It is known that the carbon–carbon bond lengths vary, thus $C_1 - C_2 = C_3 - C_4 = 0.135$ nm and $C_2 - C_3 = 0.146$ nm. This would seem to suggest that the p-orbitals must interact to a different degree across different bonds and so give rise to different resonance integrals. Let us however as a first approximation take all three resonance integrals to be equal and see how well the results justify this assumption. The secular determinant can be written down using the approximations and symbolism introduced above.

$$
\begin{array}{c c}
 & \begin{array}{c c c c} 1 & 2 & 3 & 4 \end{array} \\
\begin{array}{c} 1 \\ 2 \\ 3 \\ 4 \end{array} &
\left| \begin{array}{c c c c}
x & 1 & 0 & 0 \\
1 & x & 1 & 0 \\
0 & 1 & x & 1 \\
0 & 0 & 1 & x
\end{array} \right| = 0.
\end{array}
$$

This determinant is equivalent to

$$(x^2 - 1)^2 - x^2 = 0,$$

i.e. $(x^2 - 1 + x)(x^2 - 1 - x) = 0.$

The solutions are

$$x = \frac{-1 \pm \sqrt{(1+4)}}{2} \quad \text{or} \quad x = \frac{+1 \pm \sqrt{(1+4)}}{2},$$

i.e. $x = \dfrac{\pm 1 \pm \sqrt{5}}{2}$

$$= -1.618, \; -0.618, \; +0.618, \; +1.618.$$

The charge distribution can be found by substituting these values for x in the secular equations:

(i) $a_1 x + a_2 \qquad\qquad\quad = 0,$
(ii) $a_1 + a_2 x + a_3 \qquad\quad = 0,$
(iii) $a_2 + a_3 x + a_4 \; = 0,$
(iv) $a_3 + a_4 x = 0.$

For $x = -1.618$

(i) $-1.618a_1 + a_2 = 0,$
$$a_2 = +1.618a_1;$$

(ii) $a_1 + (1.618a_1)(-1.618) + a_3 = 0,$
$$-1.618a_1 + a_3 = 0,$$
$$a_3 = +1.618a_1;$$

(iv) $+1.618a_1 + a_4(-1.618) = 0,$
$$a_1 = +a_4.$$

Normalization:

$$a_1^2 + (1.618a_1)^2 + (1.618a_1)^2 + a_1^2 = +1.$$

Hence $\qquad\qquad 7.236a_1^2 = +1,$

i.e. $\qquad\qquad a_1 = 0.3755 = a_4$

and $\qquad\qquad a_2 = 0.6070 = a_3.$

The other coefficients may be found in the same way and summarized:

Table 3

x	a_1	a_2	a_3	a_4
-1.618	$+0.3755$	$+0.6070$	$+0.6070$	$+0.3755$
-0.618	$+0.6070$	$+0.3755$	-0.3755	-0.6070
$+0.618$	$+0.6070$	-0.3755	-0.3755	$+0.6070$
$+1.618$	$+0.3755$	-0.6070	$+0.6070$	-0.3755

To calculate charges and bond orders the following products will be useful.

Table 4

x		a_1^2	$a_1 a_2$	a_2^2	$a_2 a_3$	a_3^2	$a_3 a_4$	a_4^2
-1.618	ψ_1	$+0.141$	$+0.228$	$+0.359$	$+0.359$	$+0.359$	$+0.228$	$+0.141$
-0.618	ψ_2	$+0.359$	$+0.228$	$+0.141$	-0.141	$+0.141$	$+0.228$	$+0.359$
$+0.618$	ψ_3	$+0.359$	-0.228	$+0.141$	$+0.141$	$+0.141$	-0.228	$+0.359$
$+1.618$	ψ_4	$+0.141$	-0.228	$+0.359$	-0.359	$+0.359$	-0.228	$+0.141$

The four π molecular orbitals may be distinguished ψ_1, ψ_2, ψ_3 and ψ_4. The ground state configuration is therefore $\psi_1^2 \psi_2^2$. Charge distribution and bond orders in this and some other configurations are given in Table 5.

Table 5

	q_1	p_{12}	q_2	p_{23}	q_3	p_{34}	q_4
$\psi_1^2\psi_2^2$	1·00	0·912	1·00	0·436	1·00	0·912	1·00
$\psi_1^2\psi_2^1\psi_3^1$	1·00	0·456	1·00	0·718	1·00	0·456	1·00
$\psi_1^2\psi_2^2\psi_3^1$	1·359	0·684	1·141	0·577	1·141	0·684	1·359

It is interesting to note that whilst the charge distribution in the ground state is quite uniform over the four atoms the bond order between them is not, being much greater in the $C_1 - C_2$ and $C_3 - C_4$ regions than across the middle bond. If bond order is an indication of charge to be found between two nuclei then it is reasonable to expect some correlation between bond order and bond strength and also bond length. Thus the calculations suggest that the assumption of the same resonance integral between all the carbon $2p$ orbitals must be in error. Shorter, stronger bonds, associated with larger values of β, will be found between atoms $C_1 - C_2$ and $C_3 - C_4$ than between atoms $C_2 - C_3$. The effect of repeating the calculations using a smaller central resonance integral will be discussed later.

If an electron is excited from ψ_2 to ψ_3 or if an extra electron enters ψ_3 (as in the buta-1,3-dienyl anion) the bond-order picture is completely altered. In the excited state the central bond should now be the shortest and electrons on the terminal atoms will be relatively isolated. In the case of the anion the bond orders become much more nearly equal. This latter situation can sometimes be approached in complexes with transition metals when electrons from d-orbitals are 'back donated' to antibonding ligand π-orbitals. This simple theory indicates that in this case the bond length alternation characteristic of the free ligand should be largely destroyed (see problem 2.2).

2.2.4 π Molecular orbitals in benzene

As a final example of the working of the method let us consider benzene – the molecule upon which Hückel demonstrated the power of the molecular orbital theory and of his own simplifications in 1930. The secular determinant that corresponds to the cyclic interaction of the six carbon $2p$ orbitals is

$$
\begin{array}{c|cccccc}
 & 1 & 2 & 3 & 4 & 5 & 6 \\
\hline
1 & x & 1 & 0 & 0 & 0 & 1 \\
2 & 1 & x & 1 & 0 & 0 & 0 \\
3 & 0 & 1 & x & 1 & 0 & 0 \\
4 & 0 & 0 & 1 & x & 1 & 0 \\
5 & 0 & 0 & 0 & 1 & x & 1 \\
6 & 1 & 0 & 0 & 0 & 1 & x
\end{array} = 0.
$$

The reader may care to expand this determinant for himself and demonstrate that it corresponds to the equation

$$(x^2-4)(x^2-1)(x^2-1) = 0.$$

Fortunately the symmetry of this molecule can be used to simplify these equations (pp. 45 and 136) and the labour involved in solving six-by-six or larger determinants indicates that such simplifications would be very welcome. The energies of the six π molecular orbitals of benzene corresponding to the roots of the secular equation are $\alpha \pm 2\beta$, $\alpha \pm \beta$ and $\alpha \pm \beta$. The coefficients can be found from the secular equations in the usual way, thus for $x = -2$:

$$\psi_1 = \frac{1}{\sqrt{6}}(\phi_1 + \phi_2 + \phi_3 + \phi_4 + \phi_5 + \phi_6),$$

where ϕ_{1-6} are the six carbon $2p$ orbitals. For the degenerate pair of orbitals ψ_2 and ψ_3 ($x = -1$) we must proceed as for the cyclopropenyl system.

Let us assume in ψ_2 that the coefficient for ϕ_1 is zero, then the secular equations,

(i) $a_1 x + a_2 + \qquad\qquad\qquad a_6 = 0,$

(ii) $a_1 + a_2 x + a_3 \qquad\qquad\qquad = 0,$

(iii) $a_2 + a_3 x + a_4 \qquad\qquad = 0,$

(iv) $a_3 + a_4 x + a_5 \qquad\quad = 0,$

(v) $a_4 + a_5 x + a_6 = 0,$

(vi) $a_1 + \qquad\qquad\qquad a_5 + a_6 x = 0,$

become

(i) $0(-1) + a_2 + a_6 = 0,$
                    $a_2 = -a_6,$

(ii) $0 + a_2(-1) + a_3 = 0,$
                    $a_2 = +a_3,$

(iii) $a_2 + a_2(-1) + a_4 = 0,$
                    $a_4 = 0,$

(iv) $a_2 + 0(-1) + a_5 = 0,$
                    $a_2 = -a_5.$

With the normalizing condition we now have enough information to determine the coefficients:

$$0^2 + a_2^2 + a_2^2 + 0^2 + (-a_2)^2 + (-a_2)^2 = +1.$$

$$a_2 = \tfrac{1}{2} = a_3,$$

$$a_5 = -\tfrac{1}{2} = a_6,$$

$$a_1 = 0 = a_4.$$

Using equation **2.5** the coefficients for ψ_3 can be written down directly:

$$a_1 = +\frac{1}{\sqrt{3}} = -a_4,$$

$$a_2 = +\frac{1}{\sqrt{12}} = a_6,$$

$$a_3 = -\frac{1}{\sqrt{12}} = a_5.$$

When the π-bond orders are calculated it is found that they are the same between all the pairs of bonded atoms, justifying the equality of resonance integral assumed in the original formulation and at least suggesting that the carbon skeleton in benzene itself should be a regular hexagon.

2.2.5 π *Molecular orbitals in benzene using one element of symmetry*

It is not necessary to use the full symmetry of benzene to achieve some simplification of the 6×6 determinant. A single mirror plane perpendicular to the molecule and containing carbon atoms 1 and 4 will suffice. Since reflection in this plane is a property of the molecule it must also be a property of the charge distributions in each molecular orbital,

thus $a_{2i}^2 = a_{6i}^2$ and $a_{3i}^2 = a_{5i}^2$

for the ith molecular orbital. It therefore follows that

$$a_{2i} = \pm a_{6i} \quad \text{and} \quad a_{3i} = \pm a_{5i}.$$

We can therefore devise 'pseudo-molecular' orbitals from the atomic orbitals of carbon atoms 2 and 6, and also 3 and 5, thus

$$\psi_{i+} = \frac{1}{\sqrt{2}}(\phi_2 + \phi_6),$$

$$\psi_{ii+} = \frac{1}{\sqrt{2}}(\phi_3 + \phi_5),$$

$$\psi_{i-} = \frac{1}{\sqrt{2}}(\phi_2 - \phi_6),$$

$$\psi_{ii-} = \frac{1}{\sqrt{2}}(\phi_3 - \phi_5).$$

The $+$ and $-$ subscripts indicate whether the orbital is symmetric or antisymmetric with respect to reflection in the plane.

45   Examples of Simple Molecular Orbital Calculations

The determinant may now be rewritten:

$$
\begin{array}{c|cccc|cc}
 & \phi_1 & \psi_{i+} & \psi_{ii+} & \phi_4 & \psi_{i-} & \psi_{ii-} \\
\hline
\phi_1 & \alpha-E & \dfrac{\beta}{\sqrt{2}} & 0 & 0 & 0 & 0 \\
\psi_{i+} & \dfrac{\beta}{\sqrt{2}} & \alpha-E & \beta & 0 & 0 & 0 \\
\psi_{ii+} & 0 & \beta & \alpha-E & \dfrac{\beta}{\sqrt{2}} & 0 & 0 \\
\phi_4 & 0 & 0 & \dfrac{\beta}{\sqrt{2}} & \alpha-E & 0 & 0 \\
\hline
\psi_{i-} & 0 & 0 & 0 & 0 & \alpha-E & \beta \\
\psi_{ii-} & 0 & 0 & 0 & 0 & \beta & \alpha-E
\end{array} = 0.
$$

Thus the 6×6 determinant is reduced to a 4×4 determinant and a 2×2 determinant which may be solved in the usual way. The roots of the symmetric determinant are ± 2, ± 1 and the antisymmetric one ± 1. The introduction of the plane of symmetry also simplifies the calculation of orbital coefficients. In the antisymmetric case for example a_1 and a_4 for both orbitals must be zero, since these atomic sites contain the plane of symmetry.

Other symmetry planes could equally well have been used. The one perpendicular to the molecule and bisecting two opposite C–C bonds would for example lead to a reduction of the 6×6 determinant to two 3×3 determinants. Quite obviously symmetry properties are most valuable in reducing the labour of such calculations.

2.3 Variations of α or β

So far in these examples it has been assumed that all the Coulomb (α) and resonance (β) integrals in the calculations have been constant, although in the case of butadiene the result indicated that a somewhat smaller resonance integral should be used across the C_2–C_3 bond than across C_1–C_2 or C_3–C_4 bonds. It is quite easy to take account of such a situation; if the resonance integral for the terminal bonds is β, let that for the central bond be $k\beta$ and then solve the determinant in the usual way:

$$
\begin{array}{c|cccc}
 & 1 & 2 & 3 & 4 \\
\hline
1 & x & 1 & 0 & 0 \\
2 & 1 & x & k & 0 \\
3 & 0 & k & x & 1 \\
4 & 0 & 0 & 1 & x
\end{array} = 0,
$$

i.e. $(x^2-1)^2 - k^2 x^2 = 0,$

$$x = \pm \frac{2+k \pm \sqrt{(4k+k^2)}}{2}$$

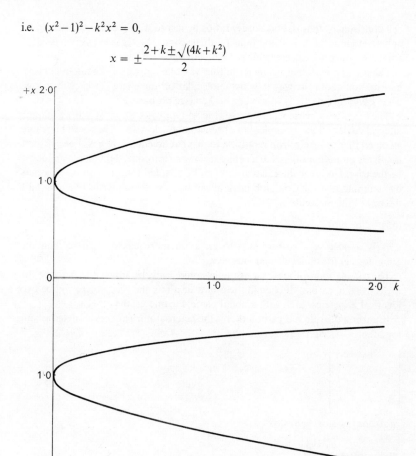

Figure 9   Variation with k of the possible values of x for the π-orbitals of the buta-1,3-diene molecule

A graph to show the variation of the possible values of x with k is shown in Figure 9. This now shows the effect of allowing two ethylenic π-orbitals to interact, until a central ethylenic bond forms with two isolated atoms at 1 and 4. Clearly with a certain amount of effort it would be possible to compute all the orbital coefficients as functions of k and so also the bond orders. This would then permit a check between k and bond order if a suitable relationship could be arrived at.

The simplest that might be tried would be

$$\frac{k}{1} = \frac{p(2,3)}{p(1,2)}.$$

47   **Variations of α or β**

Self-consistency, that is, bond-order ratios in accord with the chosen k would be only then found for $k = 0$ and buta-1,3-diene should be regarded as two isolated ethylenic systems – extreme, but not far from the truth.

The use of a variable parameter in this way often gives very useful qualitative information about the way in which molecular orbital energy levels will change when bonds are partially formed or broken (see problem 2.3).

The other basic parameter in these Hückel-type calculations is α, the Coulomb integral. So far all the examples have been homonuclear and it has been hopefully assumed that those electron repulsion terms that do enter into α will be constant for all the atoms considered. If a heteroatom were introduced this could no longer be the case. However the calculations can be adapted to deal with heteroatoms by assuming that the Coulomb integral can be expressed in terms of α and β for the rest of the molecule,

thus $\quad \alpha(\text{hetero}) = \alpha + \varepsilon\beta$.

Clearly a positive ε corresponds to an atom more electronegative than the remainder in the molecule and vice versa.

The calculations for allyl, for example, could easily be adapted to describe the π-electrons in carbon dioxide (if the atoms lie along the z-axis there will be two identical π-systems, each with a nodal plane, the one xz, the other yz).

In carbon dioxide it is carbon that is the heteroatom. The secular determinant for either π-system is

$$\begin{vmatrix} x & 1 & 0 \\ 1 & x+\varepsilon & 1 \\ 0 & 1 & x \end{vmatrix} = 0.$$

$\alpha(\text{carbon}) = \alpha(\text{oxygen}) + \varepsilon\beta$

and $\quad x = \dfrac{\alpha(\text{oxygen}) - E}{\beta}$.

The determinant reduces to

$$x(x^2 + x\varepsilon - 2) = 0,$$

thus $\quad x = 0 \quad$ or $\quad x = \dfrac{-\varepsilon \pm \sqrt{(\varepsilon^2 + 4)}}{2}$.

To proceed further it is necessary to make some rather wild and naïvely hopeful assumptions just in order to produce an answer. Let us suppose for example that β is about 3 eV and that the difference in carbon and oxygen αs is about 6 eV. (Ionization potentials for p-orbitals are little use here – they are about the same – yet oxygen is more electronegative than carbon, so say 6 eV then!) Thus ε is -2. Then the bonding orbital is at $x = -0.732$. The nonbonding $x = 0$ orbital is concentrated on the oxygen and the antibonding orbital is at $x = +2.732$ $\left(x = (E - \alpha)/\beta\right)$.

Again it is probably more instructive to plot a graph of the variation of orbital energies with ε rather than place too much faith in 'quantitative' values. The usefulness of such calculations becomes more apparent when charge distributions are found. These show, as would be expected, charge concentrating near electronegative atoms and eschewing electropositive ones. This in turn affects the electron repulsion terms which have been hopefully averaged out in this simple method. Clearly this approximation must be suspect in heteroatomic systems, so that any great attention to detail is simply a waste of time.

2.4 Electron repulsion in molecular orbitals

Of the many approximations used in the simple molecular orbital method outlined above one of the more fundamental was the assumption that electron repulsion terms could be accommodated within single-electron Hamiltonian operators. If we attempt to treat these terms explicitly we shall also have to use the determinant form of the wave function. Each individual orbital will be a molecular orbital that can be expressed in terms of component atomic orbitals by the L.C.A.O. approximation. Clearly the complexity of the resulting equations will be enormous. Approximations suggested by Pople have been used most successfully to cut the Gordian knot. The basic equations are the secular equations introduced above but the 'Coulomb' and 'resonance' integrals have a modified form. The Coulomb integral, $\alpha_r = H_{rr}$, is replaced by F_{rr}:

$$F_{rr} = h_r^{core} + \tfrac{1}{2}q_r\gamma_{rr} + \sum_{r \neq s} q_s\gamma_{rs}.$$

Similarly the resonance integral $\beta = H_{rs}$ is replaced by F_{rs}:

$$F_{rs} = \beta_{rs}^{core} - \tfrac{1}{2}p_{rs}\gamma_{rs}.$$

In these equations q_r is the charge in orbital r, p_{rs} is the bond order between orbitals r and s,

$$h_r^{core} = \int \phi_r \left[\sum_{i=1}^{n} \mathscr{H}(i)\right]\phi_r \, d\tau,$$

$$\beta_{rs}^{core} = \int \phi_r \left[\sum_{i=1}^{n} \mathscr{H}(i)\right]\phi_s \, d\tau$$

and $\quad \gamma_{rs} = \iint \phi_r(1)\phi_s(2)\left[\frac{e^2}{r(12)}\right]\phi_r(1)\phi_s(2) \, d\tau(1) \, d\tau(2).$

The expressions for F_{rr} and F_{rs} have separated out the electron repulsion terms that involve γ_{rs} and γ_{rr} from the so-called core integrals which are now the sum of true one-electron operators. Actually many electron repulsion terms have been ignored; this is the C.N.D.O. (complete neglect of differential overlap) approximation in which only repulsion terms with electron 1 in the same orbital (e.g. ϕ_r) on both sides of the $e^2/r(12)$ operator and also electron 2 in the same orbital (e.g. ϕ_s) on both sides of the operator, are included.

When applied to π-systems in organic molecules further approximations of the Hückel type can be made, that is, interactions between nonadjacent orbitals are ignored.

Let us now briefly consider the effect of these new equations upon the calculation of molecular orbital energies and charge distributions. The first step is to evaluate bond orders and charges at atoms by the simple method given above. In the case of benzene, for example, all values of q will be 1 and all values of p will be $\frac{2}{3}$. Bond lengths are all equal so γ_{rs} terms will be identical as also will the γ_{rr} terms (all the orbitals are the same). Thus for all values of r, F_{rr} will be the same, similarly for F_{rs}: $\beta_{rs}^{core} - \frac{1}{2}p_{rs}\gamma_{rs}$ will be the same across each bond. Thus constant values for F_{rr} and F_{rs} may be used, as in section 2.2.4 above. Because both bond order and charge distribution are constant, explicit consideration of electron repulsion terms has no effect.

With buta-1,3-diene the situation is a little different since, whilst all qs are $+1$, the bond orders vary considerably. Thus F_{rr} will be nearly the same at all four atoms (assuming γ_{rr} and γ_{rs} to be constant, terminal and central Fs differ only by γ_{rs}) but F_{rs} will be different at the $C_2 - C_3$ bond from its value at the terminal bonds. Now the γ-terms describe electron repulsions and will therefore be positive. β_{rs}^{core} is negative so that if we assume (as we did for buta-1,3-diene) that all the β_{rs}^{core} terms and the γ_{rs} terms are the same for the three bonds then F_{rs} will be larger for the terminal bonds than for the central bonds. Using this value of F_{rs} the calculations may be repeated and charges and bond orders calculated and compared with the original values. These new bond orders may then once again be inserted into the equation for F_{rs} to generate F'_{rs} and the calculations repeated etc. until the situation is reached when the value of p_{rs} calculated is the same (to within some defined limit of acceptable error) as that used at the beginning of that particular step of the calculation. This technique of calculation is called an iterative procedure and is ideally suitable for computers (but not for humans).

In more complicated molecules that may contain heteroatoms, or which may themselves be charged or in which for some reason or another the charges at each atom are not identical, the values of F_{rr} will also be altered by the actual charge distribution.

A very general comparison with the results of Hückel calculations is now possible. Where simple calculations indicate a disparity in charge distribution the effect of the $q_r\gamma_{rr}$ and $q_s\gamma_{rs}$ terms will be to reduce this effect. If r is electronegative q_r would tend to be greater than one (e.g. at N in pyridine) but the first calculation of F_{rr}, the core electronegativity, will be reduced by the positive γ-terms. Another simple conclusion is that differences in bond order will be enhanced, that is electron pairs will tend to become localized. This result was apparent with butadiene where the terminal F_{rs} tended to increase relative to F_{23}.

Since the hard work in these calculations is done by computers the main problem lies in choice of parameters, especially the γs. Suffice it to say that many authors have developed satisfactory recipes which give excellent results.

2.5 Molecular wave functions

So far we have concentrated our attention upon the Hamiltonian, in the first case using approximations to avoid explicit mention of electron repulsion and in the second case discussing the effect which electron repulsion terms, properly considered, might have upon the energies and charge distributions in a molecule. Let us now look at the wave functions themselves, and begin with the simplest of molecules, hydrogen. In the ground state both electrons will be in the bonding molecular orbital $\psi(i)$ with opposed spins. The over-all wave function in determinant form is therefore

$$\Psi = \frac{1}{\sqrt{2}} \begin{vmatrix} \psi(i)\alpha(1) & \psi(i)\beta(1) \\ \psi(i)\alpha(2) & \psi(i)\beta(2) \end{vmatrix},$$

i.e. $\Psi = \dfrac{1}{\sqrt{2}}[\psi(i)\alpha(1)\psi(i)\beta(2) - \psi(i)\alpha(2)\psi(i)\beta(1)]$

$$= \frac{1}{\sqrt{2}}[\psi(i)(1)\psi(i)(2)][\alpha(1)\beta(2) - \alpha(2)\beta(1)]$$

$$= \frac{1}{\sqrt{2}}\Psi_{orb}\Psi_{spin}.$$

Let us now investigate the consequences of expanding the molecular orbital as a linear combination of atomic orbitals,

i.e. $\psi(i) = \dfrac{1}{\sqrt{2}}(\phi_A + \phi_B)$

where ϕ_A and ϕ_B are the hydrogen $1s$ orbitals of the atoms A and B making up the molecule.

$$\Psi_{orb} = \tfrac{1}{2}[\phi_A(1) + \phi_B(1)][\phi_A(2) + \phi_B(2)]$$

$$= \tfrac{1}{2}[\phi_A(1)\phi_A(2) + \phi_B(1)\phi_B(2) + \phi_A(1)\phi_B(2) + \phi_B(1)\phi_A(2)].$$

The first two terms correspond to both electrons being on one of the atoms, that is, to $H_A^- H_B^+$ and $H_A^+ H_B^-$. They are called 'ionic structures'. In the last two terms electrons are shared between the two atoms, first one way round then the other – these are 'covalent structures'. Thus to use the L.C.A.O. approximation in this simple way is to imply that the bonding in hydrogen has 50 per cent ionic and 50 per cent covalent character. Intuitively this is a surprise – what should be more covalent than the bond in H_2? More detailed calculations also show that the ionic contribution is excessive. The reason is that in the simple L.C.A.O. approximation no polarization of atomic orbitals is considered: it is supposed that the orbitals are in no way altered by the proximity of other nuclei. Clearly this is incorrect but equally clearly it will be rather difficult to take account of this polarization. Other ways round the difficulty involve retaining the idea of unpolarized atomic orbitals but seek to modify Ψ_{orb} so as to reduce its ionic character.

51 Molecular Wave Functions

2.6 Valence bond method

The simplest thing to do is to ignore ionic character altogether and take as the basis for the chemical bond,

$$\Psi_{orb} = [\phi_A(1)\phi_B(2) + \phi_B(1)\phi_A(2)].$$

This approximation is used in the *valence bond* approach to chemical bond formation. Since a pair of electrons is the fundamental unit this theory treats best those molecules in which all the electrons are paired off in well-defined localized bonds. Difficulties arise with excited or ionized states and with radicals; and also in molecules such as benzene where the π-electrons can no more be thought of as localized as in Figure 10(a) than as in Figure 10(b).

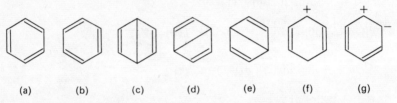

| (a) | (b) | (c) | (d) | (e) | (f) | (g) |

Figure 10   Valence bond representations of the benzene molecule: (a) and (b) Kekulé structures; (c), (d) and (e) Dewar structures; (f), (g) etc. ionic structures

To overcome this difficulty the valence bond method invokes the concept of *resonance* between the two structures (a) and (b). The word is an unfortunate choice since one tends to think of the system of π-electrons somehow swinging from (a) to (b) and back again. All that is really implied is that the over-all wave function will be a linear combination of possible pictorial structures, whose interaction will lead to a new system more stable than any of the contributing structures. Thus for a first approximation to the ground state of benzene we may write:

$$\Psi = \frac{1}{\sqrt{2}}(\psi_{(a)} + \psi_{(b)}).$$

Other structures can also be imagined (e.g. the Dewar forms (c, d, and e) and ionic structures (f, g, etc.)). If these structures are of higher energy than $\psi_{(a)}$ or $\psi_{(b)}$ their contribution to Ψ will be less. Using this approach one can 'mix in' higher and higher energy structures until a Ψ of the desired accuracy is reached.

2.7 Configuration interaction

A rather less drastic approximation than that used in the valence bond method is to try and reduce to the correct amount, rather than to eliminate, the ionic contribution. The antibonding molecular orbital for hydrogen has the form

$$\psi(ii) = \frac{1}{\sqrt{2}}(\phi_A - \phi_B).$$

Thus

$$\Psi_{orb}(\text{antibonding}) = \tfrac{1}{2}[\phi_A(1) - \phi_B(1)][\phi_A(2) - \phi_B(2)]$$

$$= \tfrac{1}{2}[\phi_A(1)\phi_A(2) + \phi_B(1)\phi_B(2) - \phi_A(1)\phi_B(2) - \phi_B(1)\phi_A(1)].$$

This is the orbital function for the doubly excited state of the hydrogen molecule. That we might imagine it corresponds to a molecule that could have no chemical stability is of no consequence : it is of the right mathematical form to modify the ionic:covalent ratio. The idea is much the same as that for resonance in the valence bond method. It is imagined that the best orbital function for the ground state may be made up from the interaction of various structures (necessarily, as we shall see in Chapter 5, all of the same symmetry) such as Ψ_{orb} and Ψ_{orb}(antibonding) introduced above. The interaction of such molecular structures produces new wave functions the most stable of which is more stable than any individual contributing structure in just the same way as in the simple theory atomic orbitals interacted to produce molecular orbitals. A good approximation to the ground state will be a structure such as Ψ_{orb} where the electrons are in the lowest energy molecular orbital. The contribution of higher-energy excited-state structures will be inversely proportional to their energy of excitation. Thus for hydrogen we would seek to express the ground-state wave function as

$$\Psi_{orb}(\text{ground}) = \Psi_{orb} - \lambda\Psi_{orb}(\text{antibonding}),$$

where λ is a constant to be found by the variation method. If the wave functions for Ψ_{orb} and Ψ_{orb}(antibonding) are examined it will be clear that the negative sign decreases ionic and increases covalent character in the ground-state wave function.

It can now be seen that both molecular orbital and valence bond methods need to include excited-state structures into their wave functions to obtain a more accurate description even of molecular ground states. This is a direct consequence of the use of the L.C.A.O. approximation. As will be discussed later in Chapter 5 the two approaches have their own conceptual advantages in different chemical problems but let us now turn to a review of the fundamentally important factors in chemical bond formation.

2.8 Conditions for bonding

There are three main conditions that must be fulfilled before chemical bonds can be formed between atoms.

2.8.1 *Orbitals must overlap*

This most reasonable assumption is made the basis of the principle of maximum overlap – that is, the strongest bonds will be formed between orbitals that overlap most. Hybrid orbitals are constructed to have enhanced directional properties and so make 'overlap' between such orbitals more directly obvious. Of course there is no real difference in final overlap in a given bond between hybrid and non-hybridized orbitals – it is just that in the latter case the contributions are fragmented. Also the principle of maximum overlap should not be used too freely since two other conditions for binding exist.

2.8.2 *Orbital energy match*

For efficient bonding the two participating orbitals should have comparable energy. This is easily demonstrated: let the Coulomb integral be

$$\alpha_A = \alpha + \varepsilon\beta$$

and $\alpha_B = \alpha - \varepsilon\beta$,

where β is the resonance integral, then the energies of the molecular orbitals can be found by solving the determinant

$$\begin{vmatrix} x+\varepsilon & 1 \\ 1 & x-\varepsilon \end{vmatrix} = 0,$$

i.e. $x^2 - \varepsilon^2 - 1 = 0$,

$$x = \pm\sqrt{(\varepsilon^2 + 1)}.$$

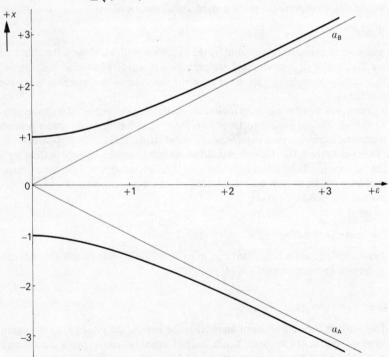

Figure 11   Variation of molecular orbital energy with ε

Figure 11 shows how the molecular orbital energies vary with ε. As ε increases the bonding molecular orbital becomes less and less bonding relative to the most electronegative atomic orbital, until in the limit (ε very large), the 'molecular' orbitals have the same energy as the atomic orbitals.

Whether bonds will be formed under such circumstances depends upon the disposition of electrons in the original atoms. Clearly if one electron were to be in orbital A and one in B there might well be energetic advantage in electron B entering the lowest molecular orbital even though it was mostly A in character and only slightly more bonding than A. Such a bond would be an ionic bond, approximately $A^- B^+$. Advantages of this type would not occur if both electrons were initially on A (see the discussion of XH_4 compounds, Chapter 5 p. 123). (i.e. Na^+Cl^- is known but the isoelectronic NeAr is not!) Thus it may be concluded that for the strongest covalent contribution to bond formation $\varepsilon \rightarrow 0$.

2.8.3 Symmetry

A third condition must be fulfilled before orbital interaction that might lead to bond formation can occur: the orbitals must have the correct symmetry. This aspect is discussed in greater detail in Chapter 5, but even the simple diagrams in Figure 12 show that in some cases although orbitals overlap effectively their overlap integral is in fact zero.

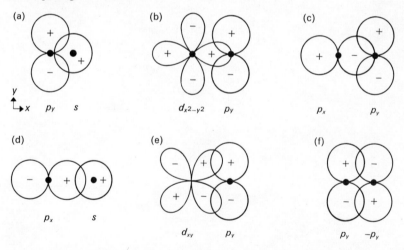

Figure 12   Some examples of orbital overlap: (a), (b) and (c) have effective overlap of zero; (d), (e) and (f) have finite (positive or negative) overlap

The symmetry properties of a molecule can be used to find out which of these orbital interactions will be zero. Since such interactions can then be ignored this results in a considerable simplification in the calculations leading to molecular orbital properties. The effectiveness of introducing just one plane of symmetry into the benzene calculations has been demonstrated. The next few chapters in this book will be devoted to a more detailed discussion of molecular symmetry properties and application to problems of chemical bonding.

Problems 2

2.1   Consider the H–H bond using the simple Hückel molecular orbital method. What is the effect of *not* neglecting the overlap integral S upon orbital energies and charge distribution?

2.2   Determine the energies (in terms of α and β) of the π molecular orbitals in (square) cyclobutadiene. Comment upon the allocation of electrons to these orbitals in the neutral molecule.

2.3   Write down the determinant for the system of three *p*-orbitals in either allyl or cyclopropenyl in which the 1–3 resonance integral can be varied. Solve the determinant and present the results graphically.

Chapter 3
Symmetry and Point Groups

Symmetry is a widely distributed property, and by no means confined to molecules. The outside of the human body is (barring minor differences) symmetric; the left-hand side is the mirror image of the right-hand side. A plane of symmetry divides the two halves. Reflection in this plane transfers the right-hand side to the left-hand side and vice versa, but the result of the operation is indistinguishable from the original. Such operations, whether they be reflections in a plane, or rotations about an axis, or inversions through a point, which generate something identical in appearance with the starting object, are *symmetry operations*.

Symmetry operations are carried out with respect to some physical feature, a plane or an axis for example. These features are called *elements of symmetry*. Thus the plane in which a reflection is performed, the axis about which a rotation takes place, are elements of symmetry.

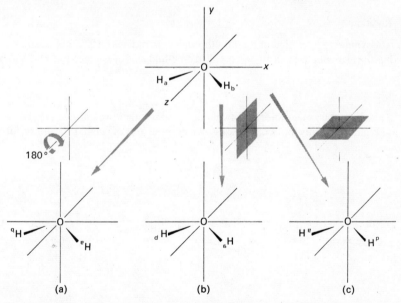

Figure 13   Symmetry operations of the water molecule: (a) rotation through 180° about the z-axis, (b) reflection in the yz plane, (c) reflection in the xz plane. Subscripts a, b have been added to show the effect of the symmetry operations

The simplest operator is the identity operator (symbol E) which leaves everything unchanged. This element of symmetry is possessed by all objects. The human body has this identity operator and reflection in the plane described above. There are no other symmetry operations. No plane of reflection divides the back from the front, nor the top from the bottom, nor is it possible to rotate the body in any way, save through 360° (which is the same as the identity operation) that will restore the body to its original appearance. An example of a rotation that is a symmetry operation is to be found in the letter S. Rotation through 180° about an axis, perpendicular to the page and passing through the middle of the letter, will reconstitute S. Rotation through 180° is therefore a symmetry operation, but rotation through any other angle between 0° and 360° is not, since the result of the operation is not identical with the original letter (for example, rotation through 90° or 270° gives ∽).

As another simple example of some symmetry operations consider the water molecule (Figure 13). Two operations (rotation through 180°, a, and a reflection, b) interchange the hydrogens and the third (another reflection, c) reflects the top of each atom into the bottom and vice versa, so that the result in each case is identical with the original molecule. These three operations and the identity operation comprise the four symmetry operations of the water molecule.

The various types of symmetry operations will now be considered in more detail.

3.1 Rotations

The symmetry operation of rotation through $2\pi/n$ radians (360°/n) about an axis has the symbol C_n. Such an axis is called an n-fold axis of symmetry.

3.1.1 C_2

The axis in the water molecule is twofold, C_2. Some other twofold axes are shown in Figure 14. Many molecules possess two or more axes of twofold symmetry, for

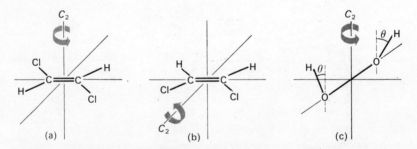

Figure 14   Examples of molecules having an axis of twofold symmetry: (a) *trans*-1,2-dichloroethylene, (b) *cis*-1,2-dichloroethylene, (c) hydrogen peroxide

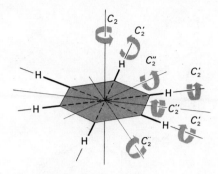

Figure 15   Twofold axes of symmetry in benzene

example benzene (Figure 15). There is a twofold axis perpendicular to the plane of the molecule which is also the site of threefold and sixfold axes of rotation, and in the plane of the molecule there are two distinct types of twofold axes, three that pass through opposite C–H bonds and three others that join the midpoints of C–C bonds. The two types of operation are distinguished by superscript dashes (', " etc.), thus C_2' and C_2''.

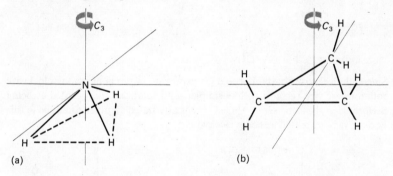

Figure 16   Examples of molecules having an axis of threefold symmetry: (a) ammonia, (b) cyclopropane

3.2.1 C_3

Figure 16 shows examples of molecules that possess a threefold axis of symmetry. Rotation about this axis may be performed in a clockwise or anticlockwise direction. Whilst either operation restores the molecule to its original appearance, neither result is identical with the original (i.e. not all the atoms are in their original places), nor are the results of either operation identical with each other (Figure 17). The clockwise and anticlockwise rotations must be regarded as two distinct examples of rotation about a threefold axis, that is C_3' and C_3'' or $\overrightarrow{C}_3$ (clockwise) and $\overleftarrow{C}_3$ (anticlockwise).

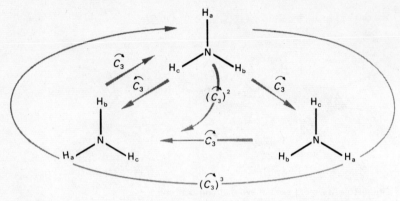

Figure 17   Results of various rotations about the C_3 axis of the ammonia molecule

If $\vec{C}_3$ is performed twice the molecule is brought into a position identical to that which results from $\overleftarrow{C}_3$ (Figure 17). This may be written as a multiplication of operations,

$$\vec{C}_3 \times \vec{C}_3 = \overleftarrow{C}_3,$$

$$\vec{C}_3^2 = \overleftarrow{C}_3.$$

Note also that $\vec{C}_3^3 = E = \overleftarrow{C}_3^3.$

Since all clockwise rotations could be expressed in terms of counterclockwise rotations, or vice versa, it will be simpler if all rotations are regarded as being performed in the same sense. Unless specifically stated all other rotations in this book may be regarded as being clockwise.

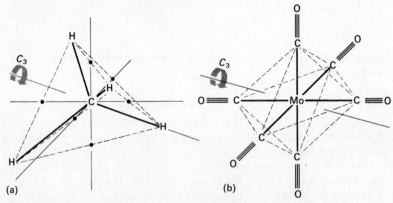

(a)   (b)

Figure 18   Examples of more complex molecules having axes of threefold symmetry: (a) methane, (b) molybdenum hexacarbonyl

More complicated examples of molecules having threefold axes of symmetry are shown in Figure 18. In methane the tetrahedron formed by joining the four hydrogens has four faces that are equilateral triangles. Each threefold axis that passes through the centre of a face and along the opposite C–H bond is the site of two symmetry operations C_3 and C_3^2.

In molybdenum hexacarbonyl, the carbon atoms lie at the vertices of a regular octahedron. Eight equilateral triangles can be formed by linking adjacent carbons. Threefold axes join the centres of opposite triangles, via the molybdenum atom, each is the location of C_3 and C_3^2 operations.

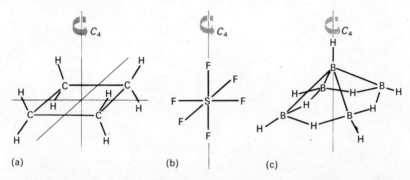

(a) (b) (c)

Figure 19   Examples of molecules containing an axis of fourfold symmetry: (a) cyclobutane – assumed flat, (b) sulphur hexafluoride, (c) pentaborane, B_5H_9

3.1.3 C_4

Figure 19 shows some molecules that contain a fourfold axis of symmetry. In cyclobutane the axis is perpendicular to the plane of the carbon atoms and passes through the centre of the molecule. In Figure 20 the carbon atoms alone are considered and it can be seen that successive C_4 operations yield four different

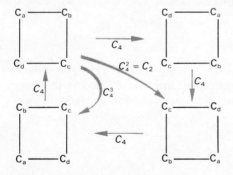

Figure 20   Effect of successive rotations of cyclobutane about the fourfold axis

symmetry operations, C_4, C_4^2, C_4^3 and C_4^4, but note that

$$C_4^2 = C_2$$

and $C_4^4 = E$.

Cyclobutane also has two other pairs of twofold axes, (a) diagonals across the carbon square and (b) joining the midpoints of the sides of the square. The operation about (a) is C_2' and about (b) is C_2''.

B_5H_9 has less total symmetry but a fourfold axis passes through the centre of the lower four boron atoms and the top boron atom.

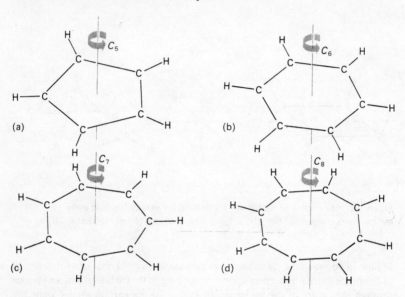

Figure 21   Molecules having five-, six-, seven- and eightfold axes of symmetry: (a) cyclopentadienyl anion, $C_5H_5^-$, fivefold, (b) benzene, C_6H_6, sixfold, (c) tropylium cation, $C_7H_7^+$, sevenfold, (d) cyclo-octatetraenyl dianion, $C_8H_8^{2-}$, eightfold

3.1.4 C_5, C_6, C_7, C_8

Examples of molecules or ions which contain such axes of symmetry are shown in Figure 21. In all cases the axis passes through the middle of the regular polygon and is perpendicular to it. Note that

$$C_6^2 = C_3, \qquad C_6^3 = C_2, \qquad\qquad C_6^4 = C_3^2,$$
$$C_8^2 = C_4, \qquad C_8^4 = C_4^2 = C_2, \qquad C_8^6 = C_4^3,$$

and that in general $C_n^n = E$ and also $C_{am}^{an} = C_m^n$.

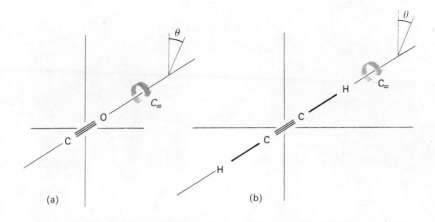

Figure 22 Examples of molecules having cylindrical symmetry: (a) carbon monoxide, (b) acetylene

3.1.5 C_∞

Molecules that have cylindrical symmetry such as acetylene, nitrogen, carbon monoxide, etc. (Figure 22) may be rotated through an indefinitely small angle θ and still the result of the rotation is identical with the starting molecule, even in the limit $\theta \rightarrow (360/\infty)°$. Such an operation is therefore given the symbol C_∞. Rotation through any arbitrary angle about such an axis is a symmetry operation.

3.1.6 *Principal axis*

In any molecule the axis of n-fold rotation C_n characterized by the largest n is the principal axis. By convention this is also made the z-axis.

3.2 **Reflections**

A plane of reflection is a two-way mirror. If such a plane can be placed in a molecule and the result of the double reflection looks the same as the original molecule then the mirror plane is an element of symmetry (see Figure 13, p. 57). Planes of reflection (symbol σ) are classified relative to the principal axis of rotation. If the plane contains the principal axis the symbol is σ_v; if the principal axis is perpendicular to the plane of reflection, symbol σ_h. A special class of σ_v is distinguished in those molecules that have twofold axes at right angles to the principal axis. If the planes of reflection bisect the angles between two adjacent twofold axes, they are given the symbol σ_d (d for dihedral).

63 Reflections

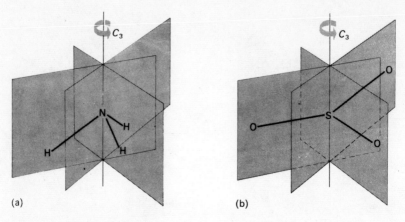

Figure 23   Mirror planes of symmetry, σ_v, in (a) ammonia and (b) sulphur
trioxide

3.2.1 σ_v

Figure 23 indicates where such planes of symmetry may be found in sulphur
trioxide and ammonia. Each of these molecules contain three σ_v planes. *Cis*-
dichloroethylene (Figure 24a) like water (Figure 13, p. 57) has two σ_v planes.

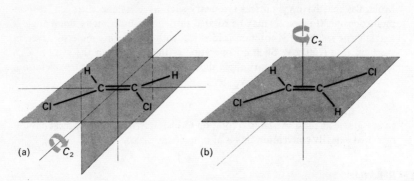

Figure 24   Planes of symmetry in dichloroethylene molecules: (a) two σ_v planes
in *cis*-dichloroethylene, (b) σ_h plane in *trans*-dichloroethylene

3.2.2 σ_h

In *trans*-dichloroethylene (Figure 24b) the plane of reflection is in the plane of
the molecule itself, at right angles to the twofold axis. Sulphur trioxide, cyclobutane and the examples shown in Figure 21 all contain, in addition to other
elements of symmetry, a σ_h plane (but note that SO_3^{2-} (isostructural with ammonia), NH_3, H_2O, B_5H_9 do *not* possess σ_h).

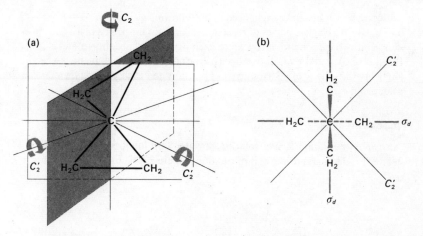

Figure 25   Spiropentane molecule, showing the σ_d planes of symmetry: (a) perspective view and (b) view along the C_2 axis

3.2.3

Two examples of molecules in which σ_d planes may be found are shown in Figure 25, spiropentane, and Figure 26, ferrocene. Ferrocene may be regarded as a ferrous ion sandwiched between two parallel (and staggered) cyclopentadienyl anions. In this molecule there are five twofold axes which pass through the iron atom and which are at right angles to the principal fivefold axis. Rotation $C_2'(1)$ (Figure 26b)

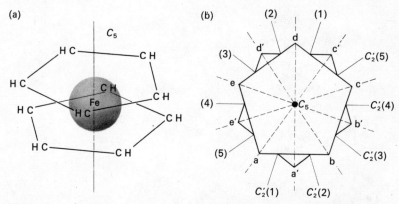

Figure 26   (a) Ferrocene, perspective view. (b) Symmetry elements of the ferrocene molecule. The fivefold axis is perpendicular to the plane of the paper, the solid lines are twofold axes and the σ_d planes contain the fivefold axis and the dashed lines

causes carbon atoms to be exchanged in the following way, a ↔ a′, b ↔ e′, c ↔ d′, d ↔ c′, e ↔ b′. The five planes of reflection all contain the principal axis, the iron atom and two carbon atoms (by which each plane is defined, thus a and c′ (1,5); b and d′ (2,3); c and e′ (4,5); d and a′ (1,2); e and b′ (3,4). Each plane bisects the angle between the twofold axes given in brackets and each plane is therefore an example of σ_d.

3.3 Improper rotations

An *improper rotation* S_n is a rotation through $2\pi/n$ followed by reflection in a plane at right angles to the axis of rotation (Figure 27). Thus

$$S_n = C_n \sigma_h.$$

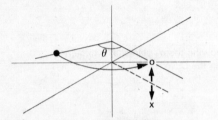

● original position of general point

○ position of ● after rotation through $\theta = \dfrac{2\pi}{n}$

× final position after reflection (◀─▶)
in plane perpendicular to axis of rotation

Figure 27   An improper rotation

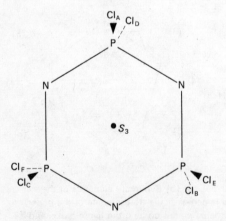

Figure 28   The $(PNCl_2)_3$ molecule, exhibiting S_3 symmetry

66   Symmetry and Point Groups

3.3.1 S_2

This is identical with inversion through a centre of symmetry and is considered by itself in the next section.

3.3.2 S_3

The molecule $(PNCl_2)_3$ (Figure 28) has this symmetry operation. The threefold axis of improper rotation is perpendicular to the plane of the paper at the centre of the equilateral triangle formed by the three phosphorus atoms. For simplicity follow the action of the S_3 operator on just one of the chlorine atoms in the molecule, Cl_A. Rotation through $\frac{1}{3} \times 360° = 120°$ brings it to the site of Cl_E. S_3 now requires reflection in a plane at right angles to the axis of rotation. If this plane contains the three phosphorus atoms the over-all effect of S_3 is to translate Cl_A to site B. At the same time all the other chlorine atoms have moved, $B \to C$, $C \to D$, $D \to E$, $E \to F$, $F \to A$. Successive operations S_3 on Cl_A will make it follow the alphabetic sequence (i.e. $S_3^2 Cl_A \to Cl_C$, $S_3^3 Cl_A \to Cl_D$ etc. until $S_3^6 Cl_A \to Cl_A$). Thus it can be seen that

$$S_3^2 = C_3^2,$$

$$S_3^3 = \sigma_h,$$

$$S_3^4 = C_3,$$

$$S_3^5 = C_3^2 \sigma_h,$$

$$S_3^6 = E.$$

These relationships can also be deduced from $S_n = C_n \sigma_h$ since $C_n^n = E$ and $\sigma_h^2 = E$.

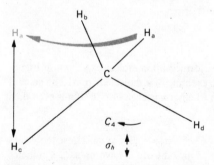

Figure 29   The S_4 symmetry operation applied to the methane molecule

3.3.3 S_4

Fourfold improper axes join the midpoints of opposite edges of a regular tetrahedron. One such axis is shown in Figure 29 in methane. S_4 first rotates H_a

through 90° until it is over, say, H_c; reflection in a plane at right angles to the axis and containing the carbon atom brings H_a to the site of H_c. S_4 causes the following simultaneous translations, $H_a \rightarrow H_c$, $H_c \rightarrow H_b$, $H_b \rightarrow H_d$, $H_d \rightarrow H_a$. Since

$$S_n = C_n \sigma_h$$

then $\quad S_4^2 = C_4^2 = C_2, \qquad S_4^3 = C_4^3 \sigma_h \quad$ and $\quad S_4^4 = E.$

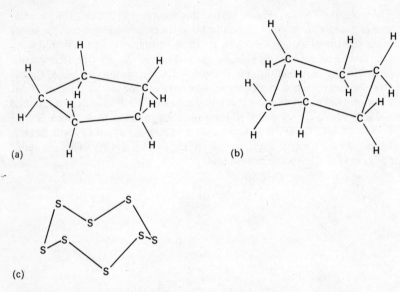

(a)

(b)

(c)

Figure 30   Examples of molecules having higher-order improper rotational symmetry: (a) cyclopentane – assumed flat – S_5, (b) cyclohexane – chair conformation – S_6, (c) sulphur $S_8 - S_8$.

3.3.4 S_5, S_6, S_8

These symmetry operations can be performed upon molecules shown in Figure 30. The sulphur molecule S_8 is a rare example of a molecular pun! The general features of the S_n operator depend on whether n is even or odd. If n is even then

$$S_n^n = C_n^n = E$$

since $\sigma_h^2 = E$ but if n is odd then $S_n^n = C_n^n \sigma_h^{n-1} \sigma_h = E E \sigma_h = \sigma_h$. Those molecules which possess an improper rotation of an odd order have σ_h as a symmetry operation. As can be seen from methane and cyclohexane (chair) this is not necessarily so when n is even. If n is even but $\frac{1}{2}n$ is odd then the molecule will also possess a centre of symmetry,

$$S_n^{\frac{1}{2}n} = C_n^{\frac{1}{2}n} \sigma_h^{\frac{1}{2}n}$$

$$= C_2 \sigma_h = S_2 = i.$$

3.4 Inversion through a centre of symmetry

A straight line is drawn from each atom to a point P, thought to be the centre of symmetry, and is then produced for an equal distance on the other side of the centre. If the act of inversion, that is to say, moving each atom through the centre to the other end of its line, reproduces the original molecule exactly then that molecule has a centre of symmetry (symbol i). In Figure 31 it can be seen that *trans*-dichloroethylene and staggered ethane both have centres of symmetry but that *cis*-dichloroethylene does not. The action of inversion causes the following translations to take place:

$$
\text{\textit{trans}-dichloroethylene} \quad
\begin{aligned}
&Cl_a \rightarrow Cl_b \quad Cl_b \rightarrow Cl_a \\
&H_c \rightarrow H_d \quad H_d \rightarrow H_c \\
&C_e \rightarrow C_f \quad C_f \rightarrow C_e,
\end{aligned}
$$

$$
\text{staggered ethane} \quad
\begin{aligned}
&C_g \rightarrow C_h \quad C_h \rightarrow C_g \\
&H_a \rightarrow H_d \quad H_d \rightarrow H_a \\
&H_b \rightarrow H_f \quad H_f \rightarrow H_b \\
&H_c \rightarrow H_e \quad H_e \rightarrow H_c.
\end{aligned}
$$

The figure also shows the result of the inversion operation on the other isomers.

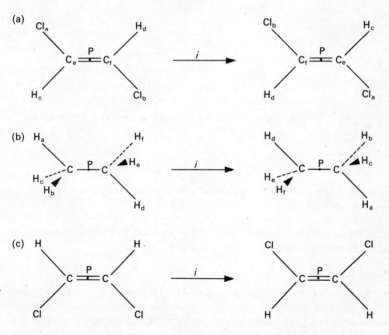

Figure 31   Inversion of (a) *trans*-dichloroethylene, (b) staggered ethane and (c) *cis*-dichloroethylene molecules through a point. (a) and (b) have centres of symmetry, (c) does not

Neither product molecule is identical with the original so that in neither is there a centre of symmetry. Other examples of molecules that do have centres of symmetry are cyclobutane (Figure 19), molybdenum hexacarbonyl (Figure 18), benzene (Figure 21), ferrocene (Figure 25), etc., and molecules that do not, ammonia (Figure 16), methane (Figure 18), B_5H_9 (Figure 19), cyclopentadienyl and the tropylium cation, cyclo-$C_7H_7^+$ (Figure 21), spiropentane (Figure 25), etc. (see problem 3.1).

3.5 Multiplication of symmetry operators

It is now possible to draw up a list of the symmetry operations possessed by any molecules. As has been shown some of the corresponding symmetry operators may be inter-related by multiplication; taking water as an example, $C_2 C_2 = E$. This idea is quite general and it is of interest to try to draw up a general multiplication table for the symmetry operators of each group to see how they are related to each other. To study the effect of multiplying two symmetry operators together it is best to consider their effect on some arbitrary point. This is shown in Figure 32, for the product $\sigma_v(xz)C_2$. The effect of the rotation C_2 upon a point

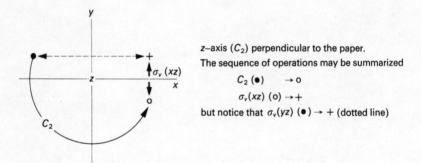

z–axis (C_2) perpendicular to the paper.
The sequence of operations may be summarized

C_2 (●) → ○

$\sigma_v(xz)$ (○) → +

but notice that $\sigma_v(yz)$ (●) → + (dotted line)

Figure 32   The multiplication of symmetry operations: $\sigma_v(xz)C_2 = \sigma_v(yz)$

followed by reflection in the xz plane takes the point to a position identical to that achieved by a simple reflection in the yz plane,

$$\sigma_v(xz)C_2 = \sigma_v(yz).$$

This is a short-hand way of writing $\sigma_v(xz) \times [C_2 \times (\text{point})] = \sigma_v(yz) \times (\text{point})$.

The operator nearest the '(point)' acts first, as symbolized by the enclosure within brackets. Thus when two operators are written multiplied together it is the right-hand operator that is taken first.

Table 6 is the complete multiplication table for the symmetry operators characteristic of the water molecule.

Table 6

second operator \ first operator	E	C_2	$\sigma_v(xz)$	$\sigma_v(yz)$
E	E	C_2	$\sigma_v(xz)$	$\sigma_v(yz)$
C_2	C_2	E	$\sigma_v(yz)$	$\sigma_v(xz)$
$\sigma_v(xz)$	$\sigma_v(xz)$	$\sigma_v(yz)$	E	C_2
$\sigma_v(yz)$	$\sigma_v(yz)$	$\sigma_v(xz)$	C_2	E

The operator considered to act first is at the top of the table, the second operator at the side and the single operator equivalent to the product is to be found where the row and column intersect (the operations in Figure 32 are indicated within the dashed lines).

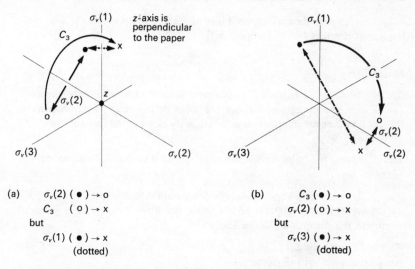

(a) $\sigma_v(2)\ (\bullet) \to o$
  $C_3\ (o) \to x$
  but
  $\sigma_v(1)\ (\bullet) \to x$
  (dotted)

(b) $C_3\ (\bullet) \to o$
  $\sigma_v(2)\ (o) \to x$
  but
  $\sigma_v(3)\ (\bullet) \to x$
  (dotted)

Figure 33 Product of C_3 and $\sigma_v(2)$ for ammonia-type symmetry: (a) $C_3\sigma_v(2) = \sigma_v(1)$, (b) $\sigma_v(2)\,C_3 = \sigma_v(3)$

Figure 33 shows the evaluation of the product of C_3 and $\sigma_v(2)$ for ammonia-type symmetry. It is most important to notice that the result depends in this case upon the order in which the symmetry operators are allowed to act,

thus $C_3\sigma_v(2) = \sigma_v(1)$ (Figure 33a)

but $\sigma_v(2)C_3 = \sigma_v(3)$. (Figure 33b)

When the product of two operators is the same, irrespective of the order they are multiplied together, then the operators are said to *commute*. Thus all the operators for water symmetry do commute, but of course the two chosen for the example from ammonia do not.

Table 7 shows the over-all multiplication table for the symmetry operators characteristic of ammonia.

Table 7

	E	C_3	C_3^2	$\sigma_v(1)$	$\sigma_v(2)$	$\sigma_v(3)$
E	E	C_3	C_3^2	$\sigma_v(1)$	$\sigma_v(2)$	$\sigma_v(3)$
C_3	C_3	C_3^2	E	$\sigma_v(3)$	$\sigma_v(1)$	$\sigma_v(2)$
C_3^2	C_3^2	E	C_3	$\sigma_v(2)$	$\sigma_v(3)$	$\sigma_v(1)$
$\sigma_v(1)$	$\sigma_v(1)$	$\sigma_v(2)$	$\sigma_v(3)$	E	C_3	C_3^2
$\sigma_v(2)$	$\sigma_v(2)$	$\sigma_v(3)$	$\sigma_v(1)$	C_3^2	E	C_3
$\sigma_v(3)$	$\sigma_v(3)$	$\sigma_v(1)$	$\sigma_v(2)$	C_3	C_3^2	E

The operators of the top row are taken first, those in the column second, thus [(column operator) × (top row operator)].

3.6 Group theory

The mathematics of group theory has been developed to deal with 'elements' (which may be symmetry operations) that obey certain rules (N.B., 'element' in this context does *not* imply or mean a symmetry element but just a member of a group).

Rule 1. A *group* is a collection of elements P, Q, etc. such that PQ is also a member of the group.

Rule 2. *Associative* (but not necessarily *commutative*) multiplication shall apply: $P(QR) = (PQ)R$. ($PQ = QP$ is not required, but if the elements of a group *do* commute, the group is said to be Abelian.)

Rule 3. The collection shall include an identity element such that $EP = PE = P$ (i.e. E commutes with all elements).

Rule 4. For every element the group shall include an inverse such that

$$QS = SQ = E.$$

If S is the inverse of Q, $S = Q^{-1}$. (An element and its inverse *do* commute.)
The multiplication tables developed above show that symmetry operators obey

Rule 1. The reader may care to verify for himself that Rule 2 also holds. Note that the symmetry operators for water form an Abelian group but that those of ammonia do not. Rule 3 is obviously obeyed. Each element does have an inverse since each column and each row does contain E. For water each operator is its own inverse but for ammonia

$$E^{-1} = E \qquad C_3^{-1} = C_3^2 \qquad (C_3^2)^{-1} = C_3,$$

$$\sigma_v(1)^{-1} = \sigma_v(1) \qquad \sigma_v(2)^{-1} = \sigma_v(2) \qquad \sigma_v(3)^{-1} = \sigma_v(3).$$

In general it can be shown that the symmetry operators possessed by any molecule do form a group as defined by Rules 1–4 above (see problem 3.2).

3.7 Classes of symmetry operators

Certain elements $(P, Q, ..., T)$ of a group are said to belong to the same *class* if

$$\left.\begin{array}{c} X^{-1}.P.X \\ X^{-1}.Q.X \\ ... \\ X^{-1}.T.X \end{array}\right\} = P \text{ or } Q \text{ or } ... \text{ or } T, \text{ but no other element,}$$

where X represents the elements of the group in turn.

3.7.1 *Example*

Consider the symmetry operators characteristic of ammonia.

Table 8

X	$X^{-1}.E.X$	$X^{-1}.C_3.X$	$X^{-1}.C_3^2.X$	$X^{-1}.\sigma_v(1).X$	$X^{-1}.\sigma_v(2).X$	$X^{-1}.\sigma_v(3).X$
E	E	C_3	C_3^2	$\sigma_v(1)$	$\sigma_v(2)$	$\sigma_v(3)$
C_3	E	C_3	C_3^2	$\sigma_v(3)$	$\sigma_v(1)$	$\sigma_v(2)$
C_3^2	E	C_3	C_3^2	$\sigma_v(2)$	$\sigma_v(3)$	$\sigma_v(1)$
$\sigma_v(1)$	E	C_3^2	C_3	$\sigma_v(1)$	$\sigma_v(3)$	$\sigma_v(2)$
$\sigma_v(2)$	E	C_3^2	C_3	$\sigma_v(3)$	$\sigma_v(2)$	$\sigma_v(1)$
$\sigma_v(3)$	E	C_3^2	C_3	$\sigma_v(2)$	$\sigma_v(1)$	$\sigma_v(3)$
	a class	a class		a class		

Thus the six symmetry operators for ammonia may be grouped into three classes. Later (p. 98) it will be shown that we need only consider the properties associated with a *representative* symmetry operator from each class, rather than with each operation individually. Symmetry operators are therefore grouped into classes, the number of operators in each class being written before the symbol for the operator, thus for the list of operators for ammonia, E, C_3, C_3^2, $\sigma_v(1)$, $\sigma_v(2)$ and $\sigma_v(3)$ can be written E, $2C_3$, $3\sigma_v$.

3.8 Simple point groups

Molecules and indeed all physical objects may be classified by the elements of symmetry they possess. Any characteristic set of symmetry operators that comprises a group defines a point group. The nomenclature for point groups depends upon the selection of important symmetry operators. The over-all symmetry of any molecule can then be described simply by the point group (rather than listing all the operators).

The various point groups can be developed from simple groups that possess only rotational symmetry by adding planes of reflection etc., and multiplying the operators together to generate all possible symmetry operators of the group.

3.8.1 C_n groups

Such groups will contain n symmetry operators in n classes, thus,

$$C_n, C_n^2, C_n^3, \ldots, C_n^n \equiv E.$$

3.8.2 C_{nh} groups

These groups will contain the n rotational operators and also the n operators in which each rotation is multiplied in turn by σ_h, thus

$$C_n, C_n^2, C_n^3, \ldots, C_n^n \equiv E, C_n \sigma_h \equiv S_n, C_n^2 \sigma_h, C_n^3 \sigma_h, \ldots, \sigma_h.$$

If n is even then the operator $C_n^{\frac{1}{2}n} = C_2$, and so a C_{nh} group would also contain $C_2 \sigma_h = S_2 = i$, a centre of inversion. If n is odd the symmetry operators of C_{nh} can be generalized, $S_n, S_n^2, S_n^3, \ldots, S_n^n \equiv \sigma_h, S_n^{n+1}, \ldots, S_n^{2n} \equiv E.$

3.8.3 C_{nv} groups

These groups are characterized by the presence of n σ_v planes of reflection as well as the n rotations of the C_n groups. When the action of any rotational symmetry operator and its inverse upon one of these σ_v planes is investigated then another plane in the same class will be generated, for example,

$$C_n^x \sigma_v C_n^{n-x} = \sigma_v^* \qquad (0 < x < n; x \text{ integral}).$$

The angle between these two planes is $(n-x)(360/n)^\circ$ (Figure 34). Thus if n is odd all n planes belong to the same class but if n is even, then two classes of planes are found each with $\frac{1}{2}n$ σ_v operators. Typical operators from each class are distinguished by a superscript dash, thus σ_v and σ_v'. The presence of σ_v planes also makes the rotations C_n^x and C_n^{n-x} members of the same class. Since $\sigma_v = \sigma_v^{-1}$, the test equations are $\sigma_v C_n^x \sigma_v = R$ and $\sigma_v R \sigma_v = C_n^x$, where the operator R will be of the same class as C_n^x. If C_n^x is a rotation through θ clockwise then R is a rotation through θ anticlockwise, thus $R = C_n^{n-x}$ (Figure 35).

In general the symmetry operators of C_{nv} groups may be summarized:

n even $\quad E, 2C_n, 2C_n^2, \ldots, C_n^{\frac{1}{2}n} \equiv C_2, \frac{1}{2}n\sigma_v, \frac{1}{2}n\sigma_v'.$

n odd $\quad E, 2C_n, 2C_n^2, \ldots, 2C_n^{\frac{1}{2}(n-1)}, n\sigma_v.$

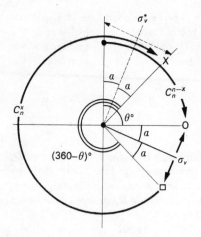

C_n^{n-x} is a rotation through $\theta°$ and C_n^x is a rotation through $(360-\theta)°$

Let the reflection O ◀━━▶ □ in the plane σ_v subtend $2a$ at the centre.

The product operation ● ━▶ O ━▶ □ ━▶ X is equivalent to a reflection in the dotted plane σ_v^* since arc ● X subtends $2a$ at the centre.

Further the angle between σ_v and σ_v^* is $\theta = (n-x)\left(\dfrac{360}{n}\right)°$

Figure 34   Generation of σ_v^* planes of symmetry in $\boldsymbol{C_{nv}}$ symmetry groups by the product $C_n^x \sigma_v C_n^{n-x} = \sigma_v^*$

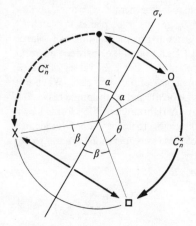

The product operation ● ━▶ O ━▶ □ ━▶ X is equivalent to an anticlockwise rotation through θ.

This will be C_n^{n-x} since C_n^x is a rotation clockwise through θ.

Figure 35   The product operator $\sigma_v C_n^x = C_n^{n-x}$

3.8.4 D_n groups

Such groups have twofold axes (operator symbol C_2') at right angles to the principal axis. The presence of these twofold axes makes C_n^x and C_n^{n-x} of the same class. The C_n operator requires $n\,C_2'$ operators of the same class whose axes are separated by an angle of $(360/n)°$. If n is even then pairs of axes will be generated that are separated by $180°$ and are coincident in space; there will be only $\frac{1}{2}n$ independent C_2' operators. However, when the product $C_n C_2'$ is investigated this generates a new C_2 operator, whose axis is perpendicular to the principal axis and which bisects the angle between any two of the existing twofold axes. By the same argument as above, there will be $\frac{1}{2}nC_2$ operators of this new type, which are given the symbol C_2''. When n is odd there are simply nC_2' operators of the same class.

n even $E, 2C_n, 2C_n^2, ..., C_n^{\frac{1}{2}n} \equiv C_2, \frac{1}{2}nC_2', \frac{1}{2}nC_2''.$

n odd $E, 2C_n, 2C_n^2, ..., 2C_n^{\frac{1}{2}(n-1)}, nC_2'.$

3.8.5 D_{nh} groups

The elements of symmetry for these groups may be derived from those for D_n groups by multiplication with σ_h. The combination of σ_h and the C_2' axes generates σ_v planes; if n is even there will be $\frac{1}{2}n\sigma_v$ and $\frac{1}{2}n\sigma_v'$ planes and if n is odd, just $n\sigma_v$ elements in one class. As with C_{nh} groups when n is odd D_{nh} groups do not have a centre of symmetry but when n is even they do.

n even $E, 2C_n, 2C_n^2, ..., C_n^{\frac{1}{2}n} \equiv C_2, \frac{1}{2}nC_2', \frac{1}{2}nC_2'', \sigma_h, 2S_n, 2C_n^2\sigma_h, ..., i, \frac{1}{2}n\sigma_v, \frac{1}{2}n\sigma_v'.$

n odd $E, 2C_n, 2C_n^2, ..., 2C_n^{\frac{1}{2}(n-1)}, nC_2', \sigma_h, 2S_n, 2S_n^3, ..., 2S_n^{n-2}, n\sigma_v.$

3.8.6 D_{nd} groups

Since the addition of σ_h to D_n groups automatically generates σ_v planes, no new groups will be formed by the addition to D_n groups of σ_v planes in which C_2' operators are located. But new groups are generated by the addition of σ_v planes that bisect the angle between two adjacent C_2' sites (n odd) or between adjacent C_2' and C_2'' sites (n even). Such planes are σ_d planes. For even values of n their presence makes the C_2' and C_2'' operators members of the same class. When n is odd the angle between adjacent C_2' sites is $(360/2n)°$ and so the angle between a σ_d plane and a C_2' operator site is $(360/4n)°$. A right angle will be formed by n such angles and so in D_{nd} groups (n odd) it is possible to find a C_2' operator located perpendicular to a σ_d plane. The product of these two operations is i and so D_{nd} groups (n odd) contain a centre of symmetry.

n even $E, 2C_n, 2C_n^2, ..., C_n^{\frac{1}{2}n} \equiv C_2, nC_2', n\sigma_d, 2S_{2n}, 2S_{2n}^3, ..., 2S_n^{n-1}.$

n odd $E, 2C_n, 2C_n^2, ..., 2C_n^{\frac{1}{2}(n-1)}, nC_2', n\sigma_d, 2S_{2n}, 2S_{2n}^3, ..., S_{2n}^n \equiv i.$

3.8.7 S_n groups

These are groups based on improper rotations, $S_n, S_n^2, ...,$ etc., if n is even then $S_n^n = C_n^n = E$ so there will be n symmetry operators but if n is odd then $S_n^n = C_n^n \times \sigma_h^n = \sigma_h$ and $S_n^{2n} = E$, and the group will have $2n$ symmetry operators. Since S_n (n odd) groups have both σ_h and $C_n^x = S_n^{2x}$ as symmetry operators these groups are identical with C_{nh} (n odd) groups and will not be considered further. For the other groups let $n = 2a$; when a is odd the S_{2a} group will have a centre of symmetry $(C_{2a}^a \sigma_h^a = C_2 \sigma_h = i)$ but if a is even it will not.

n even, $\frac{1}{2}n$ even $E, S_n, C_n, S_n^3, C_n^2, ..., C_n^{\frac{1}{2}n-1}, S_n^{n-1}.$

n even, $\frac{1}{2}n$ odd $E, S_n, C_n, S_n^3, ..., S_n^{\frac{1}{2}n} \equiv i, ..., C_n^{\frac{1}{2}n-1}, S_n^{n-1}.$

3.8.8 S_{nh} groups

In the same way as C_{nh} groups may be generated from C_n groups by the addition of a σ_h plane, so may S_{nh} groups be generated from S_n groups. When n is odd S_n already has a σ_h plane and is identical with C_{nh} so S_{nh} will also be identical with C_{nh}. When n is even the operators of an S_{nh} group will be those of S_n plus the same operators multiplied by σ_h. The operator S_n^x is the same as C_n^x when x is even but is equal to $C_n^x \sigma_h$, when x is odd. Multiplication by σ_h gives the operator $C_n^x \sigma_h^2 \equiv C_n^x$ so that in the S_{nh} group there are C_n^x operators for all values of x. The other operators of the S_{nh} group may be represented by $C_n^x \sigma_h$ so that the S_{nh} groups are identical with C_{nh} groups for all values of n.

3.8.9 S_{nv} groups

An S_{nv} group is formed by the addition of σ_v planes to an S_n group. When n is odd this is the same as adding σ_v planes to a C_{nh} group. The product $\sigma_v \sigma_h$ is equivalent to a C_2 operator acting about an axis where the planes intersect. This is a C_2' operator as found in a D_n group. Thus S_{nv} (n odd) and D_{nh} (n odd) are identical.

When n is even there are $\frac{1}{2}n$ independent σ_v planes. The product $S_n \sigma_v$ is equivalent to a rotation through $180°$ about an axis at right angles to the principal axis and bisecting the angle between adjacent σ_v planes. There will be $\frac{1}{2}n$ C_2' operators of this type. Thus the symmetry operators for S_{nv} groups (n even) are the same as those for $D_{\frac{1}{2}nd}$ groups.

3.8.10 C_{ni}, D_{ni}, S_{ni} groups

The various groups considered above have been composed from combinations of rotations and reflections. Many of these groups can also be regarded as being generated from rotational groups (C_n, D_n, S_n) and the inversion operator. In all cases the symmetry operators of these groups will be those of the rotational group plus each operator multiplied by i. The relationship of these groups to those considered above is summarized in the following table.

Table 9

i-Group		Identical to
C_{ni}	n even	C_{nh}
	n odd	S_{2n}
D_{ni}	n even	D_{nh}
	n odd	D_{nd}
S_{ni}	n even	$\frac{1}{2}n$ even C_{nh} / $\frac{1}{2}n$ odd S_{2n}
	n odd	C_{2nh}

So far all the rotation operators C_n that have been considered in any group have been through a specified angle $(360/n)°$. However, many simple molecules have an axis of cylindrical symmetry, and rotation through any arbitrary angle is a symmetry operation. The groups that contain such an axis are listed below.

Since the rotation may take place through any angle it will suffice to consider the symmetry operations associated with rotation through θ. As θ takes on different values an indefinitely large number of different operators will be generated.

Table 10

Group	Operators	Figure 22
C_∞	$E, \vec{C}_x, \overleftarrow{C}_x$	
$C_{\infty v}$	$E, 2C_x, \sigma_v$	(a)
D_∞	$E, 2C_x, C_2$	
$D_{\infty v}$	$E, 2C_x, C_2', i, 2iC_x = 2S_{2x/(x+2)}, \sigma_v$	(b)

$x = \dfrac{360}{\theta}$, i.e. C_x is a rotation through $\theta°$

$\dfrac{2x}{x+2} = \dfrac{360}{180+\theta}$, thus $S_{2x/(x+2)}$ is an improper rotation through $(180+\theta)°$.

3.9   Other groups

D_n groups were formed from C_n groups by the inclusion of twofold axes at right angles to the principal axis. Since the new axes rotate the old axis through 180° no new n-fold axes are generated. It is of some interest to inquire now what would happen if the twofold axes were not at right angles to the n-fold axis and more generally still, what if these new axes were more than twofold. The first case, with twofold axes each making an angle α to the principal axis is shown in part in Figure 36. (Actually this example will develop the second case as well.) Two of the new twofold axes are shown; there will be n such axes separated by $(360/n)°$ around the principal axis. Each axis which is the site of a C_2' operation will generate a new n-fold axis. This new axis, the C_2' site and the principal axis will also lie in a plane; two of these n planes are shown in Figure 36. Rotation about each of the new n-fold axes will generate $n-1$ more n-fold axes and $n-1$ more twofold axes; and this process could be continued indefinitely, giving rise to a vast collection of axes of no little complexity. The picture may be simplified if it is assumed that the C_n operator about the new n-fold axis (a) moves the principal axis (p) to the site of the new n-fold axis (b). In this way a random and disorganized proliferation of axes may be avoided. Furthermore, it can be shown that n can only have the values 3, 4 or 5.

Mark off unit lengths on the three n-fold axes in Figure 36

$$(OA = OB = OP = 1).$$

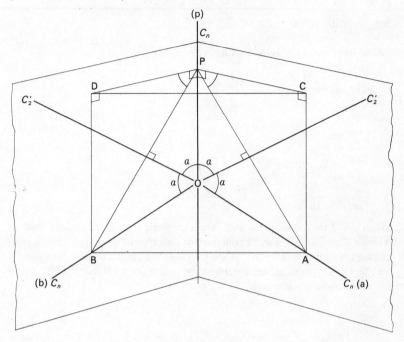

Figure 36   Generation of new symmetry axes by twofold axes at angle α to an n-fold axis

Since rotation about (p) takes (a) to (b) and rotation about (a) takes (p) to (b), $\triangle APB$ is equilateral. This implies a new threefold axis of symmetry from O, through the middle of $\triangle APB$, the corresponding operator may be given the symbol C_3'. If a plane is constructed through P at right angles to the principal axis (p), then the projection of $\angle BPA$ ($= 60°$) in that plane is $(360/n)°$. Let the projection of $\triangle APB$ be $\triangle CPD$. Triangles PAC and BPD are in planes perpendicular to $\triangle DPC$.

$$\angle BDP = \angle ACP = 90° = \angle DPO = \angle CPO.$$

Since   OA = OB = OP,

$\triangle s$ BOP and AOP are isosceles.

$$\angle DPB = \angle APC = \alpha$$

and $BP = PA = 2\sin\alpha = BA$ ($\triangle APB$ equilateral),

$$\cos\alpha = \frac{DP}{BP} = \frac{DP}{2\sin\alpha} = \frac{PC}{AP} = \frac{PC}{2\sin\alpha}.$$

Now $\quad DC^2 = DP^2 + PC^2 - 2DP.PC\cos\angle CPD,$

but $\quad\quad BA = DC = 2\sin\alpha.$

Thus $\quad 4\sin^2\alpha = 8\sin^2\alpha\cos^2\alpha - 8\sin^2\alpha\cos^2\alpha\cos\dfrac{360°}{n}$

and $\quad \cos\alpha = \left[2\left(1-\cos\dfrac{360°}{n}\right)\right]^{-\frac{1}{2}}.$

n	α
2	$60°$
3	$54°\ 46'$
4	$45°$
5	$31°\ 44'$
6	$0°$
7	negative cosine

When $n = 2$ the family of twofold axes is confined to a plane, perpendicular to which a threefold axis may be constructed. This type of group has already been considered; this example is $\boldsymbol{D_3}$. When n exceeds five there is no possible value for α so that new types of groups are only to be sought for $n = 3$ (tetrahedral), $n = 4$ (cubic) or $n = 5$ (icosahedral).

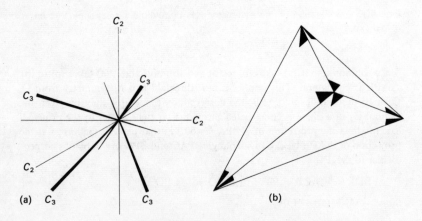

Figure 37   (a) The complete array of twofold and threefold axes of the tetrahedral group. (b) A model having tetrahedral symmetry

3.10 Tetrahedral groups

3.10.1 T

When the principal axis is threefold, the angle between like axes, 2α, is $109°$ $32'$. Figure 37(a) shows the complete array of threefold and twofold axes that characterize this group. Each of the new principal threefold axes is to be found at the site of a C'_3 operator so that there are only four threefold axes. Also the twofold axes intersect at right angles – so there can only be three independent twofold axes. The symmetry operators are E, $4C_3$, $4C_3^2$, $3C_2$. Figure 37(b) shows a model belonging to this group.

3.10.2 T_d

Planes of reflection may be added to the axes of the T-group. If the planes contain two threefold axes the C_3 operators will ensure that all six planes are of the same class. Also these planes will make C_3 and C_3^2 members of a class, $8C_3$. Each plane will also contain a twofold axis. These operators, which are characteristic of a

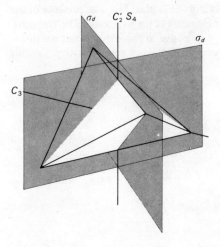

Figure 38   Planes of symmetry in the tetrahedral symmetry group

regular tetrahedron, are shown in Figure 38. By a convention the tetrahedron is usually orientated so that none of the principal threefold axes is the z-axis; rather the twofold axes are made coincident with the Cartesian axes. Also the planes of reflection are called σ_d rather than σ_v. The full complement of symmetry operators is E, $8C_3$, $6\sigma_d$, $3C_2$, $6S_4$. Methane is an example of a molecule possessing T_d symmetry (Figure 18a, p. 60).

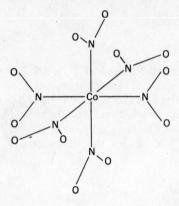

Figure 39   The hexanitritocobalt complex, which exhibits T_i symmetry

3.10.3 T_i (T_h)

If a centre of symmetry is added to T the T_i group is formed. The operators are those of T and also each of those operators multiplied by i. The presence of the centre of symmetry makes C_3 and C_3^2 members of the same class. The symmetry operators are E, $8C_3$, $3C_2$, i, $8S_6$, $3\sigma_h$ (σ_h relative to twofold axes and not three-fold). Figure 39 shows the hexanitritocobalt complex whose symmetry is of this group.

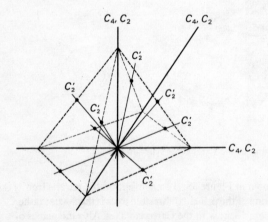

Figure 40   The family of twofold and fourfold axes of the octahedral symmetry group

3.11 Cubic or octahedral groups

3.11.1 O

Figure 40 shows the family of fourfold axes that results from the action of four twofold axes, each making 45° to the principal axis. The fourfold axes lie along Cartesian axes, so that there are only three independent fourfold axes. Also the angle between any twofold axes in the same plane is 90° so that the total number of twofold axes is only six $(6C'_2)$. The O group has other axes of twofold symmetry, coincident with the fourfold axes $3C_4^2 = 3C_2$.

Threefold axes $(C_3$ and $C_3^2)$ link the midpoints of equilateral triangles that can be constructed in opposite octants. The symmetry operators may be collected into classes in the following way, E, $6C_4$, $8C_3$, $3C_2$, $6C'_2$.

3.11.2 O_h (O_i)

This group may be generated from O by the inclusion of σ_h planes perpendicular to each fourfold axis, or (since $C_2 \sigma_h = i$) by adding a centre of symmetry. Multiplication yields the symmetry operators E, $6C_4$, $8C_3$, $3C_2$, $6C'_2$, i, $6S_4$, $8S_6$, $3\sigma_h$, $6\sigma_v$.

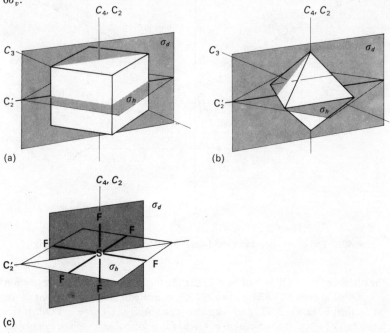

Figure 41   Some typical symmetry elements of the O_h symmetry group: (a) the cube, (b) the regular octahedron, (c) the octahedral sulphur hexafluoride molecule

The cube and the regular octahedron both have this symmetry (Figure 41). This figure also shows the location of some of the elements of symmetry in the octahedral molecule SF_6. Octahedral coordination is very widespread – for example, in SiF_6^{2-}, PCl_6^-, TeF_6, $TeCl_6^{2-}$, $PbCl_6^{2-}$, IO_6^{5-}, $Mo(CO)_6$, $Fe(CN)_6^{4-}$, etc., etc.

3.12 Icosahedral groups

3.12.1 *I* (*P*)

When $n = 5$ the five twofold axes (C_2') are each at $63°\ 28'$ to the principal axis. OA (Figure 42) is a typical new fivefold axis and about it there will stand five other fivefold axes. Two of these will be OP and OB, let OF be another, and let

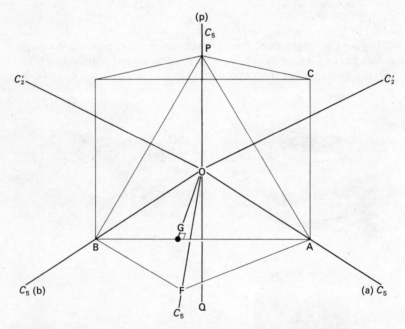

Figure 42   The generation of fivefold axes in the icosahedral symmetry group

the distance $OF = OP = OB = 1$. Extending the argument used above (section 3.9, 'Other groups'), $\triangle ABF$ is, like $\triangle ABP$, equilateral. Let the midpoint of AB be G, then F, G, O, P all lie in the same plane. Also $\angle GOP = \angle FOP$. If this angle can be found the inclination of OF to the principal axis may be calculated.

$$AP = AB = 2AG = 2\sin\alpha.$$

Since $\angle AGO = 90°$,

$$\sin \angle AOG = \frac{AG}{AO} = \frac{\sin \alpha}{1}$$

and $\quad \angle AOG = \alpha$.

Thus $\quad GO = \cos \alpha$.

In the $\triangle GOP$: $GP^2 = OP^2 + GO^2 - 2OP \cdot GO \cos \angle GOP$. Now since $\triangle ABP$ is equilateral and $AP = 2\sin\alpha$, $GP = \sqrt{3}\sin\alpha$ and this equation becomes

$$3\sin^2\alpha = 1 + \cos^2\alpha - 2\cos\alpha\cos\angle GOP,$$

thus $\quad 0 = 2\cos^2\alpha - 1 - \cos\alpha\cos\angle GOP$

and since $\alpha = 31°\ 44'$,

$$\angle GOP = 58°\ 16'.$$

Thus OF is inclined at $116°\ 32'$ to the principal axis. Let OP be projected beyond O to Q then OF makes an angle of $63°\ 28'$ with OQ. Thus the family of fivefold axes about OF will include OA, OB and OQ. Between each pair of axes will be a twofold (C_2) axis and between each adjacent trio a threefold C_3 axis. The total number of axial sites of various types is: six fivefold, ten threefold and fifteen twofold. The symmetry operators of the I (or P) group are E, $12C_5$, $12C_5^2$, $20C_3$, $15C_2$.

3.12.2 $I_i\,(I_h)$ or $P_i\,(P_h)$

The addition of planes of reflection perpendicular to the twofold (not fivefold) axes to the P-group is equivalent to the addition of a centre of symmetry. The symmetry operators are E, $12C_5$, $12C_5^2$, $15C_2$, i, $12S_{10}$, $12S_{10}^3$, $20S_6$, $15\sigma_h$. As indicated in Figure 43, the icosahedron $(B_{12}H_{12}^{2-})$ and the dodecahedron $(C_{20}H_{20}$ – not yet known), belong to this symmetry group.

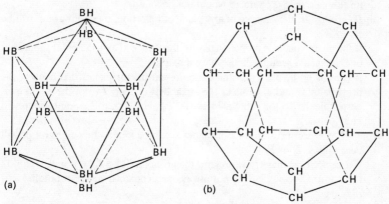

Figure 43 (a) Icosahedral anion $B_{12}H_{12}^{2-}$. (b) Dodecahedral molecule $C_{20}H_{20}$ (not yet known)

3.13 To determine the point group of a molecule

3.13.1 *Procedure*

(a) First examine the molecule to see if it has any axes of rotation. If it has none it may have a mirror plane (C_{1h}) or a centre of symmetry $(C_i \equiv S_2)$. If it has no symmetry at all the point group is C_1. If the molecule does have axes of rotation locate the axis of the highest order. This will be the principal axis. If there is more than one axis of the highest order, one of the following special cases will have been found:

three twofold axes (mutually perpendicular)   a D_{2v} group,
four threefold axis sites   a T group,
three fourfold axis sites (mutually perpendicular)   an O group,
six fivefold axis sites   a P group.

Look carefully to see if the principal axis (C_n) is also the site of an axis of improper rotation (S_{2n}).

(b) The second step is to see if there are other axes of rotation beside the principal axis. If there are none the point group will be of the C_n or S_{2n} type – depending on whether the principal axis is, or is not, also the site of an S_{2n} operator. (Proceed to (c)i.)

If there are twofold axes at right angles to the principal axis the point group will be a D_n type. (Proceed to (c)ii.)

If there are other axes of higher order than two, the molecule will belong to a point group in T, O or P (as above). (Proceed to (c)iii.)

(c) Examine the molecule for mirror planes.

(i) C_n or S_{2n} groups. If there are no mirror planes, the point group will be C_n or S_{2n}.

If the plane contains the principal axis, the group will be C_{nv} or S_{2nv}. (N.B. $S_{2nv} \equiv D_{nd}$.)

If the plane is perpendicular to the principal axis the point group will be C_{nh} (in this case an S_{2n} operator would not be found).

(ii) D_n groups. In the absence of any plane of reflection, the point group will be D_n.

If the mirror plane is perpendicular to the principal axis the point group will be D_{nh} (such a group will also have σ_vs).

If there is no σ_h plane but there are present mirror planes that contain the principal axis and also bisect the angle between any two of the n-fold axes perpendicular to the principal axis (i.e. σ_d planes) then the point group will be D_{nd}.

(iii) T, O or P groups. If no mirror planes are to be found, the point groups will be simply T, O or P.

If a centre of symmetry is present, the groups will be T_h, O_h or P_h. (N.B. In these forms the σ_h plane is not always perpendicular to the rotational axis of highest order.)

Should a T-type group have planes of reflection but no centre of symmetry the point group will be T_d.

3.13.2　*Examples*

H_2O　(a)　a twofold axis

　　　(b)　no other axes – a C_2 type group

　　　(c)i　σ_v planes – C_{2v}.

NH_3　(a)　a threefold axis

　　　(b)　no other axes – a C_3 type group

　　　(c)i　σ_v planes – C_{3v}.

S_8　(a)　a fourfold axis – also the site of an S_8 rotation

　　　(b)　twofold axes perpendicular to the principal axis join midpoints of opposite S–S bonds – D_4 type

　　　(c)ii　no σ_h but 'σ_v' planes that contain opposite S atoms. Thus these planes bisect the angle between a pair of twofold axes (b) – σ_d planes D_{4d}.

benzene　(a)　a sixfold axis

　　　(b)　twofold axes perpendicular to the sixfold axis join midpoints of opposite C–C bonds and join opposite C atoms – D_6 type

　　　(c)ii　σ_h present – D_{6h}.

spiropentane　(a)　a twofold axis – also site of S_4 operator

　　　(b)　twofold axes perpendicular to this axis, in a plane containing the central C and perpendicular to both rings; these axes bisect the angle between the planes of the two rings – a D_2 type.

　　　(c)ii　no σ_h but the planes of the rings are mirror planes; these planes bisect the angle between the twofold (C_2' type) axes – σ_d planes – D_{2d}.

(See problems 3.3 and 3.4.)

Problems 3

3.1　Draw diagrams to show the symmetry operations of the following molecules or ions:

ICl_4^-, Al_2Cl_6, P_4O_{10}, anthracene ($C_{14}H_{10}$), phenanthrene ($C_{14}H_{10}$), 1,3,5-trifluorobenzene, formaldehyde.

3.2　Deduce the symmetry operators for ethylene – do these operators form a group? – do the operators commute with each other?

3.3　What are the point groups of the examples in problem 3.1?

3.4　Consider xenon hexafluoride to have the structures:

(a) a regular octahedron,

(b) a distorted octahedron; a lone pair is located in one face pushing the three nearest fluorines away from the centre of the face. The other three fluorines approach the centre of the opposite face. (Figure 71, p. 153.)

What are the point groups for (a) and (b)? Indicate typical symmetry operations.

Chapter 4
Character Tables

An examination of a molecule will now reveal its elements of symmetry and to what symmetry group it belongs. This classification, as it stands, is of little use to the chemist. What must be done is to consider the effect of the symmetry operators, characteristic of the shape of the molecule, on the atomic or molecular orbitals that the molecule contains.

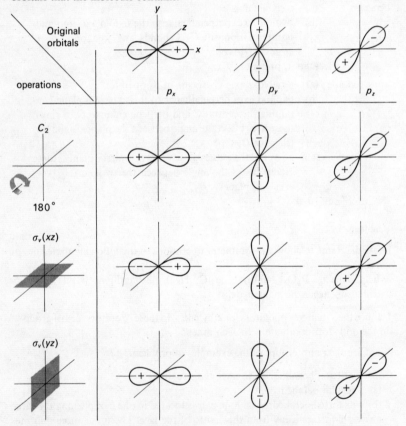

Figure 44   The effect of the operators, C_2, $\sigma_v(xz)$ and $\sigma_v(yz)$ on the three p-type orbitals

As an example of the effect of symmetry operators on atomic orbitals consider the valency orbitals of the oxygen atom in the water molecule. H_2O has C_{2v} symmetry and symmetry operators E, C_2, σ_v and σ_v'. The operator E leaves everything unchanged. Also since the $2s$ orbital of oxygen has spherical symmetry it will be unaltered by rotations or reflections which contain its centre. The effect of the other operators on the three $2p$ orbitals is shown in Figure 44. It can be seen that each operation restores each orbital to its original position, but sometimes with a change of sign. When an operator generates a new orbital identical in position and sign with the original, the effect of the operator is the same as multiplying the original orbital by $+1$. Similarly when the operator restores the orbital to its original position, but with a change of sign, this is the same as multiplying the orbital by -1. These numbers are simple examples of *matrices* each of which represents the action of an operator. The operations shown in Figure 44 can be written

$$p_y' = C_2(p_y) = (-1)p_y,$$
$$p_z'' = \sigma_v(p_z) = (+1)p_z, \quad \text{etc.}$$

For this simple example these equations may be generalized

$$\psi' = R(\psi) = [M] \times \psi,$$

where ψ and ψ' are the original and final orbitals, R is the symmetry operator and $[M]$ is the matrix that represents R.

The effect of the operators of the C_{2v} group on the oxygen orbitals can be summarized by tabulating the corresponding matrices, thus:

Table 11

	E	C_2	$\sigma_v(xz)$	$\sigma_v'(yz)$	
$2s$	$+1$	$+1$	$+1$	$+1$	(Γ_1)
$2p_x$	$+1$	-1	$+1$	-1	(Γ_2)
$2p_y$	$+1$	-1	-1	$+1$	(Γ_3)
$2p_z$	$+1$	$+1$	$+1$	$+1$	(Γ_1)

It should be observed that the multiplication properties of a row of matrices are the same as those of the operators they represent. Indeed for any suitable set of functions (orbitals, vectors etc.) there will be a collection of matrices that represent the operations of a group of symmetry operators.

Thus if $\quad PQ = R$,

then $\quad [M_P][M_Q] = [M_R]$,

where $[M_P]$ is the matrix representing action of operator P on a given orbital, or set of orbitals, and $[M_Q]$, $[M_R]$ are the corresponding matrices for operators Q, R, on the same orbital, or set of orbitals.

Any collection of matrices that obeys the multiplication laws of the symmetry operators of a group is said to form a *representation* of that group and is given the symbol Γ.

Thus the rows of matrices in Table 11 are representations of the group C_{2v}. Notice that more than one representation is possible and also that in this particular symmetry group the representation for s and p_z orbitals are the same. (This is not a general rule: the representations for each orbital must be worked out afresh for each molecule and each symmetry group.)

It is important to check that Γ_1, Γ_2, Γ_3 do in fact correspond in their multiplication to the C_{2v} operators and let us also invent a representation Γ' and see if it is a true representation.

$$\begin{array}{ccccc} & E & C_2 & \sigma_v(xz) & \sigma_v'(yz) \\ \text{e.g.} \quad \Gamma' & +1 & -1 & -1 & -1 \end{array}$$

Table 12

Operators	Γ_1	Γ_2
$EC_2 = C_2$	$+1 \times +1 = +1$	$+1 \times -1 = -1$
$E\sigma_v(xz) = \sigma_v(xz)$	$+1 \times +1 = +1$	$+1 \times +1 = +1$
$E\sigma_v'(yz) = \sigma_v'(yz)$	$+1 \times +1 = +1$	$+1 \times -1 = -1$
$C_2\sigma_v(xz) = \sigma_v'(yz)$	$+1 \times +1 = +1$	$-1 \times -1 = +1$
$C_2\sigma_v'(yz) = \sigma_v(xz)$	$+1 \times +1 = +1$	$-1 \times +1 = -1$
$\sigma_v(xz)\sigma_v'(yz) = C_2$	$+1 \times +1 = +1$	$-1 \times +1 = -1$

	Γ_3	Γ'
$EC_2 = C_2$	$+1 \times -1 = -1$	$+1 \times -1 = -1$
$E\sigma_v(xz) = \sigma_v(xz)$	$+1 \times -1 = -1$	$+1 \times -1 = -1$
$E\sigma_v'(yz) = \sigma_v'(yz)$	$+1 \times +1 = +1$	$+1 \times -1 = -1$
$C_2\sigma_v(xz) = \sigma_v'(yz)$	$-1 \times +1 = -1$	$-1 \times -1 = +1$
$C_2\sigma_v'(yz) = \sigma_v(xz)$	$-1 \times -1 = +1$	$-1 \times -1 = +1$
$\sigma_v(xz)\sigma_v'(yz) = C_2$	$-1 \times +1 = -1$	$-1 \times -1 = +1$

Thus Γ_1, Γ_2 and Γ_3 are all faithful representations of the operators of the C_{2v} group but Γ' is not (see problem 4.1).

Now let us examine the matrices that are formed when the symmetry operators are considered to act on a group of orbitals rather than just upon one at a time. Consider for example the set of valence-shell orbitals on the oxygen atom in the water molecule:

$$\begin{aligned} C_2(s) &= (+1)s + (0)p_x + (0)p_y + (0)p_z, \\ C_2(p_x) &= (0)s + (-1)p_x + (0)p_y + (0)p_z, \\ C_2(p_y) &= (0)s + (0)p_x + (-1)p_y + (0)p_z, \\ C_2(p_z) &= (0)s + (0)p_x + (0)p_y + (+1)p_z. \end{aligned}$$

This array of equations may be simplified by expressing the coefficients on the right-hand side as a four-by-four matrix, thus,

$$C_2 \begin{bmatrix} s \\ p_x \\ p_y \\ p_z \end{bmatrix} = \begin{bmatrix} +1 & 0 & 0 & 0 \\ 0 & -1 & 0 & 0 \\ 0 & 0 & -1 & 0 \\ 0 & 0 & 0 & +1 \end{bmatrix} \begin{bmatrix} s \\ p_x \\ p_y \\ p_z \end{bmatrix}.$$

Similarly, the other operators acting upon the same set of orbitals in the same order would give rise to families of equations, the coefficients in which could also be expressed as four-by-four matrices thus:

E

$$\begin{bmatrix} +1 & 0 & 0 & 0 \\ 0 & +1 & 0 & 0 \\ 0 & 0 & +1 & 0 \\ 0 & 0 & 0 & +1 \end{bmatrix},$$

$\sigma_v(xz)$

$$\begin{bmatrix} +1 & 0 & 0 & 0 \\ 0 & +1 & 0 & 0 \\ 0 & 0 & -1 & 0 \\ 0 & 0 & 0 & +1 \end{bmatrix},$$

$\sigma_v'(yz)$

$$\begin{bmatrix} +1 & 0 & 0 & 0 \\ 0 & -1 & 0 & 0 \\ 0 & 0 & +1 & 0 \\ 0 & 0 & 0 & +1 \end{bmatrix}.$$

(See problem 4.2.)

The valence orbitals of the hydrogen atoms in water could be treated in the same way. (Hydrogen atom subscripts are as in Figure 13, p. 57.)

Table 13

symmetry operation		corresponding matrix
$E \begin{bmatrix} H_a \\ H_b \end{bmatrix}$	$= (+1)H_a + (0)H_b$ $= (0)H_a + (+1)H_b$	$\begin{bmatrix} 1 & 0 \\ 0 & 1 \end{bmatrix}$
$C_2 \begin{bmatrix} H_a \\ H_b \end{bmatrix}$	$= (0)H_a + (+1)H_b$ $= (+1)H_a + (0)H_b$	$\begin{bmatrix} 0 & 1 \\ 1 & 0 \end{bmatrix}$
$\sigma_v(xz) \begin{bmatrix} H_a \\ H_b \end{bmatrix}$	$= (+1)H_a + (0)H_b$ $= (0)H_a + (+1)H_b$	$\begin{bmatrix} 1 & 0 \\ 0 & 1 \end{bmatrix}$
$\sigma_v'(yz) \begin{bmatrix} H_a \\ H_b \end{bmatrix}$	$= (0)H_a + (+1)H_b$ $= (+1)H_a + (0)H_b$	$\begin{bmatrix} 0 & 1 \\ 1 & 0 \end{bmatrix}$

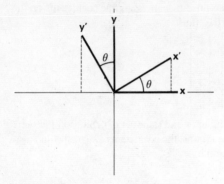

Figure 45   Effect of rotation through angle θ on a pair of orthogonal vectors
x and y

C_{2v} is a particularly simple group and all the operations can be expressed using matrices containing only 0 and ± 1. Since this also applies to the operator E it means that central atom orbitals can always be given a single representation.

Complications arise in groups with rotational symmetry operators greater than twofold. Let us consider the general case of an operator C_n that corresponds to a rotation through $\theta°$ ($n = 360/\theta$) on a pair of orthogonal vectors, for example x and y (Figure 45). Since p_x and p_y behave in this respect in the same way as x and y, the result we obtain for vectors will be directly applicable to chemical problems.

Thus $C_n \begin{bmatrix} \mathbf{x} \\ \mathbf{y} \end{bmatrix} = \begin{bmatrix} \mathbf{x}' \\ \mathbf{y}' \end{bmatrix}$,

but $\mathbf{x}' = \mathbf{x} \cos \theta + \mathbf{y} \sin \theta$

and $\mathbf{y}' = -\mathbf{x} \sin \theta + \mathbf{y} \cos \theta.$

Thus $C_n \begin{bmatrix} \mathbf{x} \\ \mathbf{y} \end{bmatrix} = \begin{bmatrix} \cos \theta & \sin \theta \\ -\sin \theta & \cos \theta \end{bmatrix} \begin{bmatrix} \mathbf{x} \\ \mathbf{y} \end{bmatrix}.$

A p_x orbital is taken completely into the position of a p_y orbital by rotation through 90°; but the $d_{x^2-y^2}$ orbital occupies completely the d_{xy} site by rotation through only 45°. For any degenerate pair of orbitals let the angle through which they must be rotated to exchange positions once be δ; δ may be called 'angle of orthogonality'.

Generalizing in the above equations we can write

$\mathbf{x}' = \mathbf{x} \cos k\theta + \mathbf{y} \sin k\theta,$

$\mathbf{y}' = -\mathbf{x} \sin k\theta + \mathbf{y} \cos k\theta,$

then $k\delta = 90°$

Table 14

	E	$C_3(\theta=120°)$	$C_3^2(\theta=240°)$	$\sigma_v(1)(xz)$	$\sigma_v(2)$	$\sigma_v(3)$
s						
p_x						
p_y						
p_z						

$E=\begin{bmatrix}1&0&0&0\\0&1&0&0\\0&0&1&0\\0&0&0&1\end{bmatrix}$

$C_3(\theta=120°)=\begin{bmatrix}1&0&0&0\\0&-\tfrac12&\tfrac{\sqrt3}{2}&0\\0&-\tfrac{\sqrt3}{2}&-\tfrac12&0\\0&0&0&1\end{bmatrix}$

$C_3^2(\theta=240°)=\begin{bmatrix}1&0&0&0\\0&-\tfrac12&-\tfrac{\sqrt3}{2}&0\\0&\tfrac{\sqrt3}{2}&-\tfrac12&0\\0&0&0&1\end{bmatrix}$

$\sigma_v(1)(xz)=\begin{bmatrix}1&0&0&0\\0&1&0&0\\0&0&-1&0\\0&0&0&1\end{bmatrix}$

$\sigma_v(2)=\begin{bmatrix}1&0&0&0\\0&\tfrac12&-\tfrac{\sqrt3}{2}&0\\0&-\tfrac{\sqrt3}{2}&-\tfrac12&0\\0&0&0&1\end{bmatrix}$

$\sigma_v(3)=\begin{bmatrix}1&0&0&0\\0&\tfrac12&\tfrac{\sqrt3}{2}&0\\0&\tfrac{\sqrt3}{2}&-\tfrac12&0\\0&0&0&1\end{bmatrix}$

$$\cos 240° = \cos 120° = -\tfrac{1}{2} \qquad \sin 120° = \frac{\sqrt{3}}{2} \qquad \sin 240° = -\frac{\sqrt{3}}{2}$$

These matrices derive from equations such as:

$$E(s) = (1)s + (0)p_x + (0)p_y + (0)p_z,$$

$$E(p_y) = (0)s + (0)p_x + (1)p_y + (0)p_z,$$

$$C_3(p_x) = (0)s + (-\tfrac12)p_x + \frac{\sqrt{3}}{2}p_y + (0)p_z,$$

$$\sigma_v(1)(p_z) = (0)s + (0)p_x + (0)p_y + (1)p_z.$$

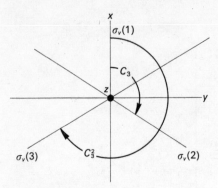

Figure 46   Symmetry operations for the C_{3v} group

and the general form of the matrix is

$$\begin{bmatrix} \cos k\theta & \sin k\theta \\ -\sin k\theta & \cos k\theta \end{bmatrix}.$$

These considerations now enable us to easily write down matrices corresponding to the action of symmetry operators for the group C_{3v} (Figure 46) acting upon the nitrogen orbitals in ammonia (see Table 14).

Thus we can see that it is possible to generate as many representations as we like depending upon the particular set of functions initially chosen. It would be most useful if there were a particular set of representations of more fundamental character. In order to see if such a set is possible it is necessary to try and reduce the size of large matrices and see if they contain these more fundamental representations.

4.1   Reduction of matrices

The actual mechanics of matrix multiplication are described in Appendix A; for this section it is sufficient to appreciate that matrices can be multiplied together. The reciprocal (or inverse) of a matrix $[X]$ can be defined as $[X]^{-1}$ and then $[X][X]^{-1} = $ 'unit matrix'. A unit matrix is one that corresponds to the symmetry operator E, all its elements are zero save along the diagonal, where they are $+1$. A matrix that has equal numbers of rows and columns is a *square matrix* (if it has n columns and n rows it is an n-by-n matrix). Let us now suppose that A, B, C, D, etc., are symmetry operators of a group represented by square matrices $[A], [B], [C], [D]$, etc. Let us also suppose

$$AB = D,$$

then $[A][B] = [D].$

New matrices can be generated by multiplication,

$$[A]' = [X]^{-1}[A][X],$$

$$[B]' = [X]^{-1}[B][X].$$

($[X]$ and $[X]^{-1}$ have the same number of rows and columns as $[A]$, $[B]$, etc.)

Then $\quad [A]'[B]' = [X]^{-1}[A][X][X]^{-1}[B][X]$

$$= [X]^{-1}[A][B][X] \quad \text{since } [X][X]^{-1} = [E]$$

$$= [X]^{-1}[D][X]$$

$$= [D]'.$$

Thus the set of matrices $[A]'$, $[B]'$, $[C]'$, $[D]'$, etc., forms a representation of the group equally as valid as $[A]$, $[B]$, $[C]$, $[D]$, etc., since it obeys the required rules of multiplication. The use of matrices $[X]$ and $[X]^{-1}$ in the above way to generate $[A]'$ etc., from $[A]$ is known as a *similarity transformation*.

Now it may be that as a result of these similarity transformations the dashed matrices have the forms,

$$[A]' = \begin{bmatrix} a_1 & 0 & 0 \\ \hline 0 & a_2 & 0 \\ \hline 0 & 0 & a_3 \end{bmatrix} \qquad [B]' = \begin{bmatrix} b_1 & 0 & 0 \\ \hline 0 & b_2 & 0 \\ \hline 0 & 0 & b_3 \end{bmatrix} \quad \text{etc.}$$

(0 represents a region of the matrix in which all the elements are zero.)

The only non-zero matrix elements occur in the regions a_1, a_2, b_1, b_2, etc., and also the submatrices with the same subscript (i.e. a_1, b_1, etc. – a_2, b_2, etc.) have the same order (i.e. have the same number of rows or columns). Following the rules of matrix multiplication for submatrices of the type $a_1 b_1$, $a_2 b_2$, etc., it can be shown that if

$$[A]'[B]' = [D]',$$

then also $\quad a_1 b_1 = d_1,$

$$a_2 b_2 = d_2, \quad \text{etc.}$$

In other words, the submatrices are also representations of the group. Thus small representations have been found within larger ones. The large matrix is said to have been *reduced*. If it is not possible to find a set of suitable matrices for the similarity transformation, the original matrix is said to be *irreducible*.

Example

Let us see if it is possible to reduce the matrices derived from the hydrogen 1s orbitals in the water molecule (p. 91).

Suppose $[x] = \begin{bmatrix} (\sqrt{2})^{-1} & -(\sqrt{2})^{-1} \\ (\sqrt{2})^{-1} & (\sqrt{2})^{-1} \end{bmatrix}$,

then $[x]^{-1} = \begin{bmatrix} (\sqrt{2})^{-1} & (\sqrt{2})^{-1} \\ -(\sqrt{2})^{-1} & (\sqrt{2})^{-1} \end{bmatrix}$.

The similarity transformation for the matrix that represents the operations of C_2 and $\sigma'_v(yz)$ on the hydrogen atoms is,

$$\begin{bmatrix} (\sqrt{2})^{-1} & -(\sqrt{2})^{-1} \\ (\sqrt{2})^{-1} & (\sqrt{2})^{-1} \end{bmatrix} \cdot \begin{bmatrix} 0 & 1 \\ 1 & 0 \end{bmatrix} \cdot \begin{bmatrix} (\sqrt{2})^{-1} & (\sqrt{2})^{-1} \\ -(\sqrt{2})^{-1} & (\sqrt{2})^{-1} \end{bmatrix}$$

$$= \begin{bmatrix} (\sqrt{2})^{-1} & -(\sqrt{2})^{-1} \\ (\sqrt{2})^{-1} & (\sqrt{2})^{-1} \end{bmatrix} \begin{bmatrix} -(\sqrt{2})^{-1} & (\sqrt{2})^{-1} \\ (\sqrt{2})^{-1} & (\sqrt{2})^{-1} \end{bmatrix}$$

$$= \begin{bmatrix} -1 & 0 \\ 0 & +1 \end{bmatrix}.$$

The matrix for the operators E and $\sigma_v(xz)$ is in a form that can obviously be reduced but for completeness let us consider that similarity transformation too.

$$\begin{bmatrix} (\sqrt{2})^{-1} & -(\sqrt{2})^{-1} \\ (\sqrt{2})^{-1} & (\sqrt{2})^{-1} \end{bmatrix} \cdot \begin{bmatrix} 1 & 0 \\ 0 & 1 \end{bmatrix} \cdot \begin{bmatrix} (\sqrt{2})^{-1} & (\sqrt{2})^{-1} \\ -(\sqrt{2})^{-1} & (\sqrt{2})^{-1} \end{bmatrix}$$

$$= \begin{bmatrix} (\sqrt{2})^{-1} & (\sqrt{2})^{-1} \\ (\sqrt{2})^{-1} & (\sqrt{2})^{-1} \end{bmatrix} \cdot \begin{bmatrix} (\sqrt{2})^{-1} & (\sqrt{2})^{-1} \\ -(\sqrt{2})^{-1} & (\sqrt{2})^{-1} \end{bmatrix}$$

$$= \begin{bmatrix} +1 & 0 \\ 0 & +1 \end{bmatrix}.$$

Thus all the matrices for the symmetry operations upon the hydrogen atoms in water can be reduced. The results of this reduction may be summarized.

E	C_2	$\sigma_v(xz)$	$\sigma'_v(yz)$	
1	-1	1	-1	(Γ_3)
1	1	1	1	(Γ_1)

Often it is possible to reduce large matrices to a collection of single elements by this method. However, when a symmetry group contains rotations of more than twofold symmetry some matrices of the type shown on p. 92 will be found when large matrices are reduced. These two-by-two matrices are irreducible unless combinations involving imaginary coefficients are formed. The component orbitals of these matrices must be considered together (a degenerate pair) under *all* the operations of this group.

4.2 The character

The sum of the diagonal elements of a square matrix (which is referred to in matrix algebra as the *spur* or the *trace*) is called the *character* and is given the symbol χ. The character has the especially useful property that it is unaltered by similarity transformations.

Proof

Let $[X]^{-1}[A][X] = [B]$.

If $[A]$ is an $n \times n$ matrix then

$$\chi_A = \sum_{i=1}^{n} a_{ii},$$

where a_{ij} is a typical matrix element.

Consider another matrix $[C]$, of the same order,

$$\chi_C = \sum_{i=1}^{n} c_{ii},$$

then the character of the product is

$$\chi_{AC} = \sum_{i=1}^{n} (ac)_{ii} = \sum_{i=1}^{n} \sum_{l=1}^{n} a_{il} c_{li}.$$

Since the matrix elements are scalar numbers the order of their multiplication and summation will not matter.

$$\chi_{AC} = \sum_{i=1}^{n} \sum_{l=1}^{n} c_{li} a_{il} = \sum_{i=1}^{n} (ca)_{ii} = \chi_{CA}.$$

Thus the character of the product matrix $[AC]$ is the same, whichever is the order of multiplication.

Hence $\quad \chi_B = \chi_{X^{-1}AX} = \chi_{X^{-1}XA} = \chi_{EA} = \chi_A.$

So if $[A]$ and $[B]$ are matrices related by a similarity transformation their character will be the same.

Since large matrices may be reduced and so shown to contain irreducible components by similarity transformations alone it follows that

$$\chi_R = \sum_{i=1}^{n} a_i \chi_i(R),$$

where χ_R is the character of a large matrix for the Rth operation of a group, $\chi_i(R)$ is the character of the ith irreducible representation for the Rth operation and a_i is the number of times $\chi_i(R)$ is contained in χ_R.

The irreducible components of a large matrix can thus be found using the character alone without having to find the matrices $[X]$ and $[X]^{-1}$ each time. This obviously will reduce the labour involved enormously. The task will be made even simpler if all the values of $\chi_i(R)$ are also known.

4.3 Character tables for point groups

The tabulation of the values of $\chi_i(R)$ for all i irreducible representations for each of the symmetry classes of a group is a character table for that particular point group. A collection of character tables is given at the end of this book.

The symmetry elements of the same *class* are related by a similarity transformation. Thus all the matrices in a given representation, that correspond to symmetry operators of the same class have the same character. So, if only the character is to be found, it need be evaluated for only one representative symmetry operator in each class. This means that for many purposes the symmetry operators of a group may be grouped together in their classes, thus for the group C_{3v}, $-E$, C_3, C_3^2, $\sigma_v(1)$, $\sigma_v(2)$ and $\sigma_v(3)$ become E, $2C_3$, $3\sigma_v$, the number of elements per class being written before the symmetry operator.

The irreducible representations are distinguished by a symbolism that depends on their symmetry properties (see Table 15).

Table 15

A, B	singlet	$\chi(E) = 1$
$E\dagger$	doubly degenerate representation	$\chi(E) = 2$
T (or F)	triply degenerate representation	$\chi(E) = 3$
A	symmetric ⎫ with respect to rotation	$\chi(C_2) = +1$
	through 180° about	
B	antisymmetric⎭ principal axis	$\chi(C_2) = -1$
Superscripts $\left\{\begin{array}{l} ' \\ '' \end{array}\right.$	symmetric ⎫ reflection in σ_h	$\chi(\sigma_h) = +1$
	antisymmetric⎭	$\chi(\sigma_h) = -1$
Subscripts $\left\{\begin{array}{l} g \\ u \end{array}\right.$	symmetric (*gerade*) ⎫ inversion in a centre	$\chi(i) = +1$
	antisymmetric (*ungerade*)⎭ of symmetry	$\chi(i) = -1$
1	symmetric ⎫ reflection in σ_v	$\chi(\sigma_v) = +1$
2	antisymmetric⎭	$\chi(\sigma_v) = -1$

† Not to be confused with E the identity operation.

Numerical subscripts $(1, 2, 3, \ldots, n)$ are also used in a general way to distinguish representations and do not always indicate symmetry relative to σ_v.

Lower-case letters (a, b, e, t) are used for the symmetries of atomic and molecular orbitals; capital letters for the symmetries of over-all atomic or molecular states.

4.4 Direct product

It is often necessary to multiply together the representations of two sets of functions, such as $A_1, A_2, \ldots, A_i$, denoted by $\Gamma(A)$, and $B_1, B_2, \ldots, B_j$, denoted by $\Gamma(B)$. The action of a representative operator R may be expressed

$$R(A_i) = \sum_j a_{ji} A_j$$

and $$R(B_k) = \sum_l b_{lk} B_l.$$

Then the operation of R upon the product would be

$$R(A_i B_k) = \sum_j \sum_l a_{ji} b_{lk} A_j B_l.$$

The product functions must also form a representation of the group so that it is possible to write

$$R(A_i B_k) = \sum_j \sum_l c_{jl,ik}(A_j B_l). \qquad \textbf{4.1}$$

Now the character is derived from the sum of the diagonal elements of the matrix,

thus $\chi_{R(AB)} = \sum_i \sum_k c_{ik,ik}(A_i B_k).$

Using equation **4.1** (since $i = j$ and $k = l$)

$$\chi_{R(AB)} = \sum_i \sum_k a_{ii} b_{kk}(A_i B_k)$$

$$= \chi_{R(A)} \chi_{R(B)}.$$

Thus the direct product character, for a given operator R is simply the product of the characters in the representation of A and B for the same operation.

$$\chi_{R(\text{direct product})} = \chi_{R(A)} \chi_{R(B)}.$$

4.5 Reduction of large matrices using character tables

The tabulation of irreducible representations (I.R.s) facilitates the reduction of large matrices since the character of the large matrix is the sum of the characters of the irreducible matrices into which it may be transformed.

$$\chi(R) = \sum_i a_i \chi_i(R).$$

Multiply by $\chi_j(R)$, and sum over all operators R:

$$\sum_R \chi(R) \chi_j(R) = \sum_R \sum_i a_i \chi_i(R) \chi_j(R).$$

This equation may be simplified utilizing the property of orthogonality possessed by irreducible representations,

i.e. $\sum_R \chi_i(R) \chi_j(R) = h \delta_{ij},$

where h is the number of symmetry operations in the group, $\chi_i(R)$ is the character of ith I.R. for operation R and δ_{ij} is known as the Kronecker delta ($\delta_{ij} = +1$ for $i = j$; $\delta_{ij} = 0$ for $i \neq j$).

Thus $\sum_R \chi_i(R) \chi_j(R)$ is only non-zero for $i = j$.

Hence $\sum_R \chi(R) \chi_j(R) = a_j h$

and
$$a_j = \frac{1}{h} \sum_g n_R \chi(R) \chi_j(R),$$

where n_R is the number of symmetry operations in the class of R and g is the number of classes in the group

(i.e. $h = \sum_{R=1}^{g} n_R$).

To find the various values of a_j requires a knowledge of $\chi_j(R)$; these values are tabulated in character tables (Appendix B).

Examples

(a) The representation for the two hydrogen atoms in water $\Gamma(2H)$ is

$$
\begin{array}{ccccc}
& E & C_2 & \sigma_v(xz) & \sigma_v'(yz) \\
\Gamma(2H) & \begin{bmatrix} 1 & 0 \\ 0 & 1 \end{bmatrix} & \begin{bmatrix} 0 & 1 \\ 1 & 0 \end{bmatrix} & \begin{bmatrix} 1 & 0 \\ 0 & 1 \end{bmatrix} & \begin{bmatrix} 0 & 1 \\ 1 & 0 \end{bmatrix}
\end{array}
$$

Characters are   2   0   2   0

a_j is the number of times the jth irreducible representation occurs, thus

$$
\begin{array}{cccc}
E & C_2 & \sigma_v(xz) & \sigma_v'(yz)
\end{array}
$$

$a(A_1) = \frac{1}{4}(1 \times 1 \times 2 + 1 \times 1 \times 0 \quad +1 \times 1 \times 2 \quad +1 \times 1 \times 0) \quad = +1,$

$a(A_2) = \frac{1}{4}(1 \times 1 \times 2 + 1 \times 1 \times 0 \quad +1 \times -1 \times 2 + 1 \times -1 \times 0) = 0,$

$a(B_1) = \frac{1}{4}(1 \times 1 \times 2 + 1 \times -1 \times 0 + 1 \times 1 \times 2 \quad +1 \times -1 \times 0) = +1,$

$a(B_2) = \frac{1}{4}(1 \times 1 \times 2 + 1 \times -1 \times 0 + 1 \times -1 \times 2 + 1 \times 1 \times 0) \quad = 0.$

Thus $\Gamma(2H) = \Gamma(A_1) + \Gamma(B_1).$

(See problem 4.3.)

(b) For the three hydrogen atoms in ammonia (C_{3v})

$$
\begin{array}{cccc}
& E & C_3 & \sigma_v \\
\Gamma(3H) & \begin{bmatrix} 1 & 0 & 0 \\ 0 & 1 & 0 \\ 0 & 0 & 1 \end{bmatrix} & \begin{bmatrix} 0 & 1 & 0 \\ 0 & 0 & 1 \\ 1 & 0 & 0 \end{bmatrix} & \begin{bmatrix} 1 & 0 & 0 \\ 0 & 0 & 1 \\ 0 & 1 & 0 \end{bmatrix}
\end{array}
$$

χs   3   0   1

$a(A_1) = \frac{1}{6}(1 \times 1 \times 3 \quad +2 \times 1 \times 0 \quad +3 \times 1 \times 1) \quad = +1,$

$a(A_2) = \frac{1}{6}(1 \times 1 \times 3 \quad +2 \times 1 \times 0 \quad +3 \times -1 \times 1) = 0,$

$a(E) = \frac{1}{6}(1 \times 2 \times 3 \quad +2 \times -1 \times 0 + 3 \times 0 \times 1) \quad = +1.$

$\Gamma(3H) = \Gamma(A_1) + \Gamma(E).$

(c) For the p-orbitals of the nitrogen in ammonia:

$$
\begin{array}{ccc}
E & C_3 & \sigma_v \text{ (e.g. } xz) \\
\end{array}
$$

$$
\begin{array}{c}
p_x \\ p_y \\ p_z
\end{array}
\begin{bmatrix}
1 & 0 & 0 \\
0 & 1 & 0 \\
0 & 0 & 1
\end{bmatrix}
\begin{bmatrix}
\cos 120° & \sin 120° & 0 \\
-\sin 120° & \cos 120° & 0 \\
0 & 0 & 1
\end{bmatrix}
\begin{bmatrix}
1 & 0 & 0 \\
0 & -1 & 0 \\
0 & 0 & 1
\end{bmatrix}
$$

clearly these matrices are reducible, and we do not need to consult the character tables.

	E	C_3	σ_v
$\chi s(p_x, p_y)$	2	$2\cos 120° = -1$	0
$\chi s(p_z)$	1	1	1

Thus $\quad \Gamma(p_x, p_y) = E$

and $\quad \Gamma(p_z) = A_1$.

In practice there is usually no need to write out the full matrix representation for each problem: it is sufficient to be able to write down the character. The following rule will be of great importance: if a symmetry operator completely moves a function away from its original position this will correspond to a zero on the diagonal of the matrix – it will not contribute to the over-all character.

This is because the diagonal elements of the matrix describe how much of an orbital (or function) remains in its original position after the operation. Obviously, if the orbital is moved completely away there will be a zero on the diagonal and thus a zero contribution to the over-all character.

It will be seen that with this idea in mind the characters for the representation $\Gamma(2H)$ and $\Gamma(3H)$ can be written down directly.

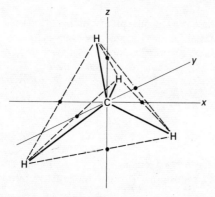

Figure 47   Orientation of the methane molecule with respect to the Cartesian axes. In this position the Cartesian axes are sites of C_2 and S_4 symmetry operations

101   Reduction of Large Matrices Using Character Tables

(d) Consider a more complicated example CH_4 (methane) T_d.

The molecule is orientated so that the x-, y- and z-axes join midpoints of opposite edges of the tetrahedron made by the four hydrogen atoms. These axes are sites of C_2 and S_4 symmetry operators. Threefold axes lie along C–H bonds. σ_d planes contain two C–H bonds (Figure 47 and also Figure 38, p. 81).

The characters for the four hydrogen atoms will be

operator	result	χ
E	all still	4
C_2	all move	0
S_4	all move	0
σ_d	2 move, 2 still	2
C_3	3 move, 1 still	1

$\Gamma(4H) = A_1 + T_2$.

For the p-orbitals on carbon,

operator	result			χ
E	all still			3
C_2 (e.g. z)	p_z still $(+1)$	$p_x \rightarrow -p_x(-1)$	$p_y \rightarrow -p_y(-1)$	-1
S_4 (e.g. z)	$p_z \rightarrow -p_z(-1)$	$p_x \rightarrow p_y(0)$	$p_y \rightarrow p_x(0)$	-1
σ_d (e.g. contains z)	p_z still $(+1)$	$p_x \rightarrow p_y(0)$	$p_y \rightarrow p_x(0)$	$+1$
C_3 (e.g. C to H atom with coordinates $(+x, +y, +z)$)	$p_x \rightarrow p_y(0)$	$p_y \rightarrow p_z(0)$	$p_z \rightarrow p_x(0)$	0

Γ (p-orbitals) $= T_2$

(See problem 4.4.)

4.6 Irreducible-representation wave functions

It is important now to decide what these irreducible representations correspond to. We now know that the valency orbitals of the two hydrogens in water taken together belong to two different irreducible representations. This may be understood if new wave functions are developed, linear combinations of the old,

thus $\quad \psi_i = \dfrac{1}{\sqrt{2}}(H_a + H_b)$

and $\quad \psi_{ii} = \dfrac{1}{\sqrt{2}}(H_a - H_b)$.

($1/\sqrt{2}$ is the normalizing factor, with neglect of overlap.)

The symmetry properties of the new orbitals ψ_i and ψ_{ii} may now be determined,

	E	C_2	$\sigma_v(xz)$	$\sigma'_v(yz)$		
ψ_i	1	1	1	1	A_1	(Γ_1)
ψ_{ii}	1	-1	1	-1	B_1	(Γ_3)

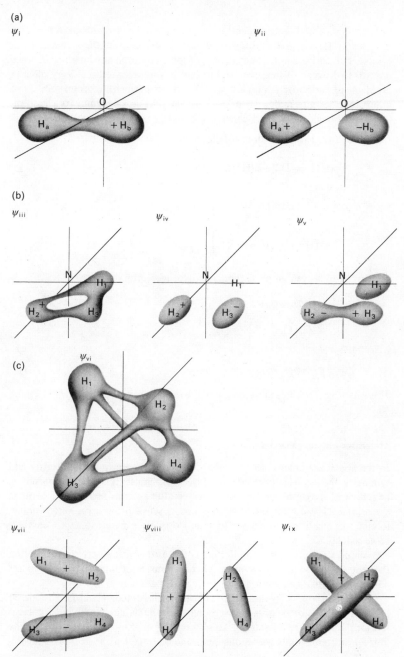

Figure 48   Irreducible representations of the hydrogen orbitals in (a) water, ψ_i and ψ_{ii}; (b) ammonia, ψ_{iii}, ψ_{iv} and ψ_v; (c) methane, ψ_{vi}, ψ_{vii}, ψ_{viii} and ψ_{ix}

These new orbitals independently correspond to the two irreducible representations for $\Gamma(2H)$. Since ψ_i and ψ_{ii} cover more than one atom they are molecular orbitals but notice that it is not necessary for there to be any bonding between the two hydrogens. (There *may* be, but that is another question.) Very often in problems of molecular bonding it will be most convenient to group orbitals or atoms together into molecular orbitals for the purposes of symmetry classification even though the orbitals concerned may be many Ångström units apart.

For the three hydrogen atoms of ammonia:

$$\psi_{iii} = \frac{1}{\sqrt{3}}[\phi(H1) + \phi(H2) + \phi(H3)], \qquad A_1$$

$$\left. \begin{array}{l} \psi_{iv} = \dfrac{1}{\sqrt{2}}[\phi(H2) - \phi(H3)], \\[2ex] \psi_v = \dfrac{1}{\sqrt{6}}[2\phi(H1) - \phi(H2) - \phi(H3)], \end{array} \right\} \quad E$$

and for the hydrogen atoms in methane (hydrogen orbitals, $\phi_1, \ldots, \phi_4$):

$$\psi_{vi} = \tfrac{1}{2}(\phi_1 + \phi_2 + \phi_3 + \phi_4), \qquad A_1$$

$$\left. \begin{array}{l} \psi_{vii} = \tfrac{1}{2}(\phi_1 + \phi_2 - \phi_3 - \phi_4), \\[1ex] \psi_{viii} = \tfrac{1}{2}(\phi_1 - \phi_2 + \phi_3 - \phi_4), \\[1ex] \psi_{ix} = \tfrac{1}{2}(\phi_1 - \phi_2 - \phi_3 + \phi_4). \end{array} \right\} \quad T_2$$

These hydrogen orbitals in water, ammonia and methane are illustrated in Figure 48.

4.7 Molecules with no central atom

In the above examples a set of orbitals has been chosen rather arbitrarily and symmetry operations performed with this set. Obviously this set of orbitals must be sufficiently large so that the symmetry operations merely regroup or rearrange the orbitals. The optimum will be to choose a set of orbitals just large enough to fulfil this condition: to include too many orbitals is merely to have a cumbersome reduction.

A simple distinction in the molecules considered above is between orbitals on the central atom and orbitals on peripheral atoms. Sometimes molecules do not have an atom that can conveniently be regarded as central.

For the purpose of symmetry classification all orbitals could be considered together but it is often possible to group the orbitals into small, but sufficient sets for the symmetry operations. To see how this might work let us consider the four $2p$ carbon orbitals perpendicular to the molecular plane in *trans*-buta-1,3-diene (C_{2h}).

$$
\begin{array}{c}
\text{C}_4 \\
\| \\
\text{C}_2-\text{C}_3 \\
\| \\
\text{C}_1
\end{array}
$$

Consider the representation for these symmetry operations (note the order of the atoms in the matrix):

$$
\begin{array}{c}
 \\
\Gamma \\
\end{array}
\begin{array}{c}
\text{C}_1 \\
\text{C}_4 \\
\text{C}_2 \\
\text{C}_3 \\
\end{array}
\quad
\overset{E}{
\begin{bmatrix}
1 & 0 & 0 & 0 \\
0 & 1 & 0 & 0 \\
0 & 0 & 1 & 0 \\
0 & 0 & 0 & 1 \\
\end{bmatrix}}
\overset{C_2}{
\begin{bmatrix}
0 & 1 & 0 & 0 \\
1 & 0 & 0 & 0 \\
0 & 0 & 0 & 1 \\
0 & 0 & 1 & 0 \\
\end{bmatrix}}
\overset{i}{
\begin{bmatrix}
0 & -1 & 0 & 0 \\
-1 & 0 & 0 & 0 \\
0 & 0 & 0 & -1 \\
0 & 0 & -1 & 0 \\
\end{bmatrix}}
\overset{\sigma_h}{
\begin{bmatrix}
-1 & 0 & 0 & 0 \\
0 & -1 & 0 & 0 \\
0 & 0 & -1 & 0 \\
0 & 0 & 0 & -1 \\
\end{bmatrix}}
$$

It is now immediately apparent that suitable matrices could have been composed using orbitals of atoms C_1 and C_4 as a pair and also orbitals at C_2 and C_3 as another pair. There was no need to write down 4×4 matrices, two rows of 2×2 would have been sufficient. This is, of course, just a process of reducing matrices visually before you start. Characters could as easily be written down directly, thus,

$$\Gamma(\text{C}_1, \text{C}_4) \quad 2 \quad 0 \quad 0 \quad -2 \quad B_g + A_u$$
$$\Gamma(\text{C}_2, \text{C}_3) \quad 2 \quad 0 \quad 0 \quad -2 \quad B_g + A_u$$

This subgrouping tells us that C_1 and C_4 together generate B_g and A_u molecular orbitals and that C_2 and C_3 together do the same. This is much more useful than merely knowing that all four orbitals would somehow combine to form two B_g m.o.s and two A_u m.o.s (see problem 4.5).

4.8 Representations of molecular orbitals

Molecular orbitals may be treated by the same methods that have been developed for atomic orbitals. By contrast with the ideas described for atomic orbitals it is very necessary to remember that molecular orbitals are spread out over the whole molecule and are not concentrated near a particular atom.

4.8.1 *Example: the π molecular orbitals of buta-1,3-diene*

The four π molecular orbitals are as shown below when derived using the Hückel approximations (Chapter 1) and making the assumption that all the resonance integrals are the same.

$$\psi_1 = 0.36\phi_1 + 0.61\phi_2 + 0.61\phi_3 + 0.36\phi_4,$$

$$\psi_2 = 0.61\phi_1 + 0.36\phi_2 - 0.36\phi_3 - 0.61\phi_4,$$

$$\psi_3 = 0.61\phi_1 - 0.36\phi_2 - 0.36\phi_3 + 0.61\phi_4,$$

$$\psi_4 = 0.36\phi_1 - 0.61\phi_2 + 0.61\phi_3 - 0.36\phi_4.$$

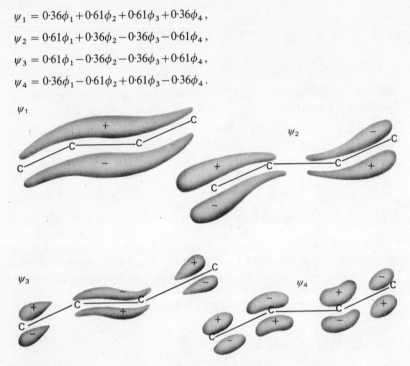

Figure 49   The π molecular orbitals of buta-1,3-diene

These orbitals are shown diagrammatically in Figure 49, and the following symmetry properties may be verified by inspection.

Table 16

	E	C_2	i	σ_h	
$\Gamma(\psi_1)$	$+1$	$+1$	-1	-1	A_u
$\Gamma(\psi_2)$	$+1$	-1	$+1$	-1	B_g
$\Gamma(\psi_3)$	$+1$	$+1$	-1	-1	A_u
$\Gamma(\psi_4)$	$+1$	-1	$+1$	-1	B_g

(See problem 4.6)

The methods outlined in this and the preceding chapters now permit the classification of molecules by their symmetry properties and the classification of the orbitals in these molecules as corresponding to various irreducible representations. The use of this scheme of classification in simplifying chemical problems will be elaborated in the remaining chapters of this book.

Problems 4

4.1 Is

	E	C_2	$\sigma_v(yz)$	$\sigma_v'(xz)$
Γ	1	1	1	-1

a true representation in the point group C_{2v}?

4.2 H_2S has the same symmetry as H_2O. Write down matrices which describe the action of the C_{2v} symmetry operators upon the five $3d$ orbitals of sulphur.

4.3 Find the characters of each of the matrices from problem 4.2. Reduce the representation. Do any of the d-orbitals belong to the same representations as the hydrogen orbitals?

4.4 Determine the point group for diborane (B_2H_6) and write down the representations for the following sets of orbitals:
(i) bridge hydrogens,   (iii) boron $2p$ orbitals,
(ii) terminal hydrogens,   (iv) boron $2s$ orbitals.
Reduce the representations.

4.5 Repeat the calculations outlined on p. 105 for *cis*-buta-1,3-diene.

4.6 Classify the molecular orbitals of the allyl system (Figure 7, p. 34) according to their irreducible representations.

Chapter 5
Chemical Bonding

The usefulness of symmetry concepts is apparent when integrals of the type $\int \phi_a A \phi_b \, d\tau$ have to be solved. When $A = \mathcal{H}$, the Hamiltonian operator, the integral is one we have already met in Chapter 1 in the evaluation of molecular orbital energy levels using normalized atomic functions. In Chapter 1, a great many approximations had to be introduced in order to solve the Schrödinger equation even for small molecules. Despite these approximations the equations were often quite complicated. Using the symmetry of the molecule, however, it is possible to effect a simplification of the Schrödinger equation without introducing any further approximations at all.

Consider $\quad E = \int \phi_a \mathcal{H} \phi_b \, d\tau.$ $\qquad\qquad$ **5.1**

This equation must have a symmetry balance just as it must balance in dimensions or units,

thus $\quad \Gamma(E) = \Gamma(\phi_a) \times \Gamma(\mathcal{H}) \times \Gamma(\phi_b).$ $\qquad\qquad$ **5.2**

Now E is a scalar number, and so it should remain the same in magnitude and sign under any symmetry operator of any group. Thus the character that corresponds to the action of any operator on the energy E must be always $+1$, and so $\Gamma(E)$ will be a row of $+1$s in any group. All groups have such a representation, which is said to be 'wholly symmetric' (e.g. A_1 in C_{2v}, A_{1g} in O_h etc.).

The representation for the Hamiltonian operator will also be wholly symmetric since, if ψ is an eigenfunction of $\mathcal{H}$, then $\mathcal{H}\psi = E\psi$ and also $\Gamma(\mathcal{H})\Gamma(\psi) = \Gamma(E)\Gamma(\psi)$. $\Gamma(\mathcal{H})$ must be the same as $\Gamma(E)$ – wholly symmetric. Thus the characters of the direct product of $\Gamma(\phi_a)$ and $\Gamma(\phi_b)$ on the right-hand side of equation 5.2 must be multiplied by the characters for $\Gamma(\mathcal{H})$, which will be $+1$ in every case. Equation **5.2** therefore reduces to

$\Gamma(\text{wholly symmetric}) = \Gamma(\phi_a)\Gamma(\phi_b).$

Now the direct product of two representations (both with $\chi(E) = +1$) will only yield the wholly symmetric representation if those two representations are the same. In the case of representations where the character for the identity operation is greater than $+1$ the direct product will be reducible but one of the irreducible components will be wholly symmetric only if the original representations were identical.

Thus equation **5.1** can only yield non-zero values of E when ϕ_a and ϕ_b belong to the same irreducible representation. Conversely E will necessarily be zero when ϕ_a and ϕ_b have different representations. The secular determinant which arises from the use of the Hückel molecular orbital method can therefore be considerably simplified if the original atomic orbitals are arranged so as to correspond to irreducible representations of the molecule. Where rows and columns corresponding to the orbitals of different representations interact there will be a zero. Usually in symmetric molecules a great many zeros are generated in this way and large determinants are easily reduced to smaller ones.

Examples

Water (C_{2v}). In the water molecule it is possible to construct orbitals from the hydrogen 1s atomic orbitals that correspond to representations a_1 and b_1 (see p. 102). The valence-shell orbitals of oxygen have I.R.s (irreducible representations): $s-a$, p_z-a_1, p_x-b_1, p_y-b_2 (xz is the plane of the molecule). Thus we need only consider the interaction of the hydrogens' a_1 orbital with the oxygen s and p_z orbitals and separately the interaction of p_x with the b_1 hydrogen orbital. The p_y orbital will not enter into the bonding at all. The determinants to be solved are thus considerably reduced from one 6×6 determinant to a 3×3, a 2×2 and a 1×1 determinant. The overall determinant is shown below, the reduction to smaller components being indicated with dotted lines.

$$
\begin{array}{c}
\overbrace{\hspace{6em}}^{a_1} \quad \overbrace{\hspace{4em}}^{b_1} \quad \overbrace{\hspace{2em}}^{b_2} \\
\begin{array}{ccccccc}
& O(s) & O(p_z) & \psi_1 & O(p_x) & \psi_2 & O(p_y) \\
a_1 \left\{ \begin{array}{c} \\ \\ \\ \end{array} \right. &
\left| \begin{array}{ccc} \alpha_s-E & 0 & \beta_1 \\ 0 & \alpha_p-E & \beta_2 \\ \beta_1 & \beta_2 & \alpha_H-E \end{array} \right.
& & & \left. \begin{array}{cc} 0 & 0 \\ 0 & 0 \\ 0 & 0 \end{array} \right|
& & \left. \begin{array}{c} 0 \\ 0 \\ 0 \end{array} \right| = 0
\end{array}
\end{array}
$$

$$
\begin{vmatrix}
\alpha_s-E & 0 & \beta_1 & 0 & 0 & 0 \\
0 & \alpha_p-E & \beta_2 & 0 & 0 & 0 \\
\beta_1 & \beta_2 & \alpha_H-E & 0 & 0 & 0 \\
0 & 0 & 0 & \alpha_p-E & \beta_3 & 0 \\
0 & 0 & 0 & \beta_3 & \alpha_H-E & 0 \\
0 & 0 & 0 & 0 & 0 & \alpha_p-E
\end{vmatrix} = 0
$$

$$\psi_1 = \frac{1}{\sqrt{2}}(H_a+H_b)\,a_1, \qquad \psi_2 = \frac{1}{\sqrt{2}}(H_a-H_b)\,b_1,$$

where α_s is the Coulomb integral for the oxygen 2s orbital, α_p is the Coulomb integral for the oxygen 2p orbitals and α_H is the Coulomb integral for the hydrogen 1s orbitals (and therefore also for ψ_1 and ψ_2 assuming no interaction between H_a and H_b); β_1, β_2, β_3 are the resonance integrals.

The three small determinants are

$$
\begin{array}{ccc}
a_1 & b_1 & b_2 \\
\begin{vmatrix} \alpha_s-E & 0 & \beta_1 \\ 0 & \alpha_p-E & \beta_2 \\ \beta_1 & \beta_2 & \alpha_H-E \end{vmatrix} = 0 &
\begin{vmatrix} \alpha_p-E & \beta_3 \\ \beta_3 & \alpha_H-E \end{vmatrix} = 0 &
\begin{vmatrix} \alpha_p-E \end{vmatrix} = 0.
\end{array}
$$

If we now assume that the s-orbital of oxygen is tightly bound and plays

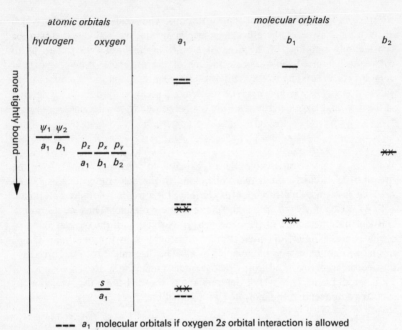

--- a_1 molecular orbitals if oxygen 2s orbital interaction is allowed

Figure 50   Energy-level diagram for the molecular orbitals of the water molecule

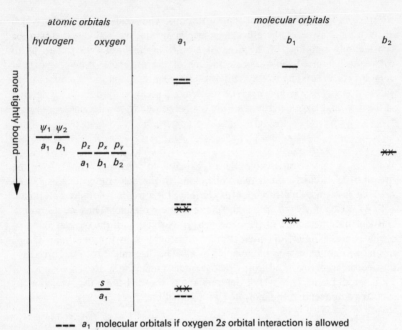

Figure 51   Energy-level diagram for the molecular orbitals of the ammonia molecule

relatively little part in the bonding the a_1 determinant is further reduced. The eight valency electrons in the water molecule are to be found in (i) a bonding orbital of a_1 symmetry, (ii) a bonding orbital of b_1 symmetry, (iii) a lone pair of b_2 symmetry and (iv) a lone pair in the oxygen $2s$ orbital of a_1 symmetry. These molecular orbitals are shown diagrammatically in Figure 50.

Ammonia (C_{3v}). A similar approach may be used for ammonia. The nitrogen orbitals have symmetries: $s - a_1$, $p_z - a_1$, p_x and $p_y - e$. The three hydrogen orbitals may be combined together (p. 104) to yield orbitals corresponding to I.R.s a_1 and e. The former will interact only with the nitrogen s and p_z orbitals, the latter pair only with p_x and p_y. Now ψ_{iv} and ψ_v are orthogonal and it can easily be arranged that ψ_{iv}, say, has the same symmetry properties as p_x and that ψ_v corresponds to p_y. This permits a further simplification within e-symmetry since ψ_{iv} will now be orthogonal to p_y and ψ_v orthogonal to p_x. The over-all determinant is

$$
a_1 \left\{
\begin{array}{ccc|cc|cc}
 & \overbrace{\hspace{3.5cm}}^{a_1} & & & \multicolumn{3}{c}{\overbrace{\hspace{5cm}}^{e}} \\
N(s) & N(p_z) & \psi_{iii} & N(p_x) & \psi_{iv} & N(p_y) & \psi_v \\
\alpha_s - E & 0 & \beta_1 & 0 & 0 & 0 & 0 \\
0 & \alpha_p - E & \beta_2 & 0 & 0 & 0 & 0 \\
\beta_1 & \beta_2 & \alpha_H - E & 0 & 0 & 0 & 0 \\
\hline
0 & 0 & 0 & \alpha_p - E & \beta_3 & 0 & 0 \\
0 & 0 & 0 & \beta_3 & \alpha_H - E & 0 & 0 \\
\hline
0 & 0 & 0 & 0 & 0 & \alpha_p - E & \beta_3 \\
0 & 0 & 0 & 0 & 0 & \beta_3 & \alpha_H - E \\
\end{array}
\right. = 0.
$$

The usual meanings are given to the αs and βs. In Figure 51 the bonding molecular orbitals are shown diagrammatically.

5.1 Hybridization

These examples show how useful symmetry ideas are in simplifying the calculation of molecular orbital energies. However, from the point of view of chemistry, the price has been high – the simple concept of the chemical bond has been lost. In the molecular orbitals described above electron pairs were delocalized over three or four nuclei and not concentrated between just two nuclei. If the charge distributions in each molecular orbital were worked out and then added up in the regions where 'bonds' were expected, it would be found that the total charge density would be about two, that is the charge on an electron pair. Thus the localized and delocalized pictures are equivalent – merely two ways of representing the same reality.

But the localized electron-pair bond is a most important chemical concept so let us see if it is possible to rearrange atomic orbitals to form new orbitals which have enhanced directional properties. Each new orbital would point to just one ligand

and form a localized bond with it. Such orbitals may be made by taking linear combinations of conventional atomic orbitals, and their formation is called *hybridization*.

Symmetry properties can be used to decide which atomic orbitals should be combined together to form a set of hybrids with a given directional character. Since localized bonds are to be formed between ligand atoms and a central atom the number and symmetry of the ligands will determine the number and juxtaposition of the hybrids. The representation for the positions of the ligands in space must be determined; the atomic orbitals that must be used in combination to form suitable hybrids will have the same representation. The simplest hybrids will be equivalent to each other, differing only in orientation. Hybrid orbitals should be normalized and orthogonal to each other.

These conditions permit the atomic orbital coefficients in each hybrid to be calculated.

Examples

sp Hybridization. Let us assume a linear system such as N_3^- or CO_2 and consider just σ-bonds from the central atom. In both cases the point group is $D_{\infty h}$. Ligand σ-orbitals will have symmetry $a_{1g} + a_{2u}$, and so will the s-orbitals and p_z orbitals on the central atom. Hybrids could therefore be

$$\psi_1 = as + bp_z,$$

$$\psi_2 = cs + dp_z,$$

with the normalizing conditions

$$a^2 + b^2 = +1 = c^2 + d^2$$

and $\quad a^2 + c^2 = +1 = b^2 + d^2.$

Hence $\quad b^2 = c^2,$

$$a^2 = d^2.$$

The orthogonality condition may be expressed as $ac + bd = 0,$

whence $\quad a = \pm d$

and $\quad b = \mp c.$

For the equivalence of the hybrid orbitals, consider the σ_h operation:

$$\sigma_h \begin{bmatrix} \psi_1 \\ \psi_2 \end{bmatrix} = \begin{bmatrix} 0 & 1 \\ 1 & 0 \end{bmatrix} \begin{bmatrix} \psi_1 \\ \psi_2 \end{bmatrix}.$$

Substituting for ψ_1 and ψ_2 in the equation for $\sigma_h(\psi_1)$,

$$\sigma_h(+ds - cp_z) = ds + cp_z = 0(ds - cp_z) + 1(cs + dp_z).$$

Equating now the coefficients for s and for p_z,

$$d = c \quad \text{(for } s\text{-orbitals)}$$

and $\quad c = d \quad$ (for p_z orbitals).

The wave functions of the sp hybrids are

$$\psi_1 = \frac{1}{\sqrt{2}}s - \frac{1}{\sqrt{2}}p_z,$$

$$\psi_2 = \frac{1}{\sqrt{2}}s + \frac{1}{\sqrt{2}}p_z.$$

sp^2 *Hybridization*. Suppose three equivalent orbitals are to be formed. Three points in a plane in the point group $\boldsymbol{D_{3h}}$ belong to representations a_1' and e'; so do the orbitals s and p_x, p_y. Hybrids will therefore be

$$\psi_1 = as + bp_x + cp_y,$$

$$\psi_2 = ds + ep_x + fp_y,$$

$$\psi_3 = gs + hp_x + ip_y.$$

The normalization requirements are

$$a^2 + b^2 + c^2 = d^2 + e^2 + f^2 = g^2 + h^2 + i^2 = +1,$$

also $\quad a^2 + d^2 + g^2 = b^2 + e^2 + h^2 = c^2 + f^2 + i^2 = +1.$

The orthogonality condition is

$$ad + be + cf = ag + bh + ci = dg + eh + fi = 0.$$

The equivalence of the orbitals can be represented by

$$C_3 \begin{bmatrix} \psi_1 \\ \psi_2 \\ \psi_3 \end{bmatrix} = \begin{bmatrix} 0 & 1 & 0 \\ 0 & 0 & 1 \\ 1 & 0 & 0 \end{bmatrix} \begin{bmatrix} \psi_1 \\ \psi_2 \\ \psi_3 \end{bmatrix}.$$

Now the result of rotating a p_x orbital through an angle θ can be expressed $p_x \cos\theta + p_y \sin\theta$ and for p_y: $\quad -p_x \sin\theta + p_y \cos\theta$.

For the equation $C_3, \theta = 120°: \cos 120° = -\frac{1}{2} = +\frac{1}{2}\sqrt{3}.$

Thus $\quad C_3(\psi_1) = \psi_2,$

which can be written as

$$as + b\left[-\frac{1}{2}p_x + \frac{\sqrt{3}}{2}p_y\right] + c\left[-\frac{\sqrt{3}}{2}p_x - \frac{1}{2}p_y\right] = as - \frac{1}{2}(b + c\sqrt{3})p_x + \frac{1}{2}(b\sqrt{3} - c)p_y$$

$$= ds + ep_x + fp_y.$$

Also $C_3(\psi_2) = \psi_3,$

hence

$$ds + e\left[-\frac{1}{2}p_x + \frac{\sqrt{3}}{2}p_y\right] + f\left[-\frac{\sqrt{3}}{2}p_x - \frac{1}{2}p_y\right] = ds - \tfrac{1}{2}(e + f\sqrt{3})p_x + \tfrac{1}{2}(e\sqrt{3} - f)p_y$$

$$= gs + hp_x + ip_y.$$

Equating coefficients for the s, p_x and p_y orbitals respectively:

(s-orbital) $a = d = g.$

Thus $a = \dfrac{1}{\sqrt{3}}.$

(p_x orbital) $h = -\dfrac{e}{2} - \dfrac{f\sqrt{3}}{2}$

and $e = -\dfrac{b}{2} - \dfrac{c\sqrt{3}}{2}.$

(p_y orbital) $i = \dfrac{e\sqrt{3}}{2} - \dfrac{f}{2}$

and $f = \dfrac{b\sqrt{3}}{2} - \dfrac{c}{2}.$

Now p_x and p_y form a degenerate pair so it is not possible to define the coefficients in a unique way. The simplest solution is to make an arbitrary assumption for one coefficient and then evaluate the others. For example assume ψ_1 is orientated along the x-axis, so that $c = 0$. Then, since $a = 1/\sqrt{3}$ and ψ_1 is normalized, $b = \sqrt{\tfrac{2}{3}}$.

Consequently $e = -\dfrac{1}{2}\dfrac{\sqrt{2}}{3} = -\dfrac{1}{\sqrt{6}}$ and $f = \dfrac{1}{\sqrt{2}}.$

Also $h = +\dfrac{1}{2}\dfrac{1}{\sqrt{6}}\left(-\dfrac{1}{2}\right)\dfrac{\sqrt{3}}{2} = -\dfrac{1}{\sqrt{6}}$ and $i = -\dfrac{1}{\sqrt{6}}\dfrac{\sqrt{3}}{2} - \dfrac{1}{2\sqrt{2}} = \dfrac{1}{\sqrt{2}}.$

Summarizing:

$$\psi_1 = \frac{1}{\sqrt{6}}(s\sqrt{2} + 2p_x),$$

$$\psi_2 = \frac{1}{\sqrt{6}}(s\sqrt{2} - p_x + p_y\sqrt{3}),$$

$$\psi_3 = \frac{1}{\sqrt{6}}(s\sqrt{2} - p_x - p_y\sqrt{3}).$$

In the same way it can be shown that four equivalent hybrid orbitals pointing to the four corners of a regular tetrahedron T_d can be composed from one s and three p-orbitals (or s and d_{xz}, d_{yz}, d_{xy}), thus:

$$\psi_1 = \tfrac{1}{2}(s + p_x + p_y + p_z),$$
$$\psi_2 = \tfrac{1}{2}(s + p_x - p_y - p_z),$$
$$\psi_3 = \tfrac{1}{2}(s - p_x + p_y - p_z),$$
$$\psi_4 = \tfrac{1}{2}(s - p_x - p_y + p_z).$$

(See problem 5.1.)

This method can be extended and hybrid orbitals constructed for any conceivable arrangement of ligands provided the atomic orbitals of the corresponding representations are present.

Since hybrid orbitals are constructed orthogonally they provide a useful picture of the regions of space that a set of electrons of equal energy and spin would occupy relative to each other at any instant. If as an approximation we assume all valence-shell electrons to be of about the same energy (often they are not), then the most probable relative disposition of four electrons of the same spin (for example) would be at the vertices of a regular tetrahedron.

Electron repulsion effects and ligand nuclei repulsions also are minimized when molecules assume the shapes suggested by equivalent hybrid orbitals. Since the simple molecular orbital approach rather crudely attempts to average out these effects, it is often an advantage, in such calculations, to assume a molecular shape based on hybridization.

If all the orbitals mixed in a hybrid are of the same energy then interhybrid interactions will be zero. Unfortunately this is rarely so and integrals of the type $\int \psi_1 \mathscr{H} \psi_2 \, d\tau$ (ψ_1, ψ_2: hybrids) may sometimes be comparable in magnitude to true resonance integrals. By way of example consider methane using sp^3 hybrids on the central carbon (principal interactions assumed between H atoms and hybrids with the same subscript).

ψ_1	H_1	ψ_2	H_2	ψ_3	H_3	ψ_4	H_4	
$\alpha_C - E$	β_1	β_2	β_3	β_2	β_3	β_2	β_3	$= 0$
β_1	$\alpha_H - E$	β_3	0	β_3	0	β_3	0	
β_2	β_3	$\alpha_C - E$	β_1	β_2	β_3	β_2	β_3	
β_3	0	β_1	$\alpha_H - E$	β_3	0	β_3	0	
β_2	β_3	β_2	β_3	$\alpha_C - E$	β_1	β_2	β_3	
β_3	0	β_3	0	β_1	$\alpha_H - E$	β_3	0	
β_2	β_3	β_2	β_3	β_2	β_3	$\alpha_C - E$	β_1	
β_3	0	β_3	0	β_3	0	β_1	$\alpha_H - E$	

α_C = Coulomb integral for sp^3 carbon hybrid.
α_H = Coulomb integral for hydrogen.

β_1 = resonance integral in a C–H bond.

β_3 = resonance integral between hybrid ψ_a and H_b $(a \neq b)$.

β_2 = 'resonance' integral between two sp^3 hybrids.

e.g.
$$\begin{aligned}
\beta_2 &= \int \psi_1 \mathcal{H} \psi_2 \, d\tau \\
&= \tfrac{1}{4} \int (s + p_x + p_y + p_z) \mathcal{H} (s + p_x - p_y - p_z) \, d\tau \\
&= \tfrac{1}{4} \left(\int s \mathcal{H} s \, d\tau + \int p_x \mathcal{H} p_x \, d\tau - \int p_y \mathcal{H} p_y \, d\tau - \int p_z \mathcal{H} p_z \, d\tau \right) \\
&= \tfrac{1}{4}(\alpha_s - \alpha_p).
\end{aligned}$$

α_s = Coulomb integral for s-orbital.

α_p = Coulomb integral for p-orbital.

Thus β_2 will be small if the energy difference in ionization potentials between s and p atomic orbitals is small compared with β_1 but when this difference is large, β_2 will no longer be negligible. It seems more reasonable to suppose that β_3 might well be small. If it is possible to set β_2 and β_3 equal to zero the determinant reduces to four identical 2×2 determinants whose roots yield the energies of the bonding and antibonding orbitals of the localized C–H bonds. In molecules where it is not reasonable to ignore β_2, calculations based on the use of hybrid orbitals will be increasingly in error. Fortunately for organic chemists, $\alpha_s - \alpha_p$ for carbon is only about -4 eV (i.e. $\beta_2 \simeq -1$ eV) whilst β_1 for many important bonds to carbon (e.g. C–C, C–H) is about 5 eV. Thus $\beta_2/\beta_1 \simeq 0.2$ which may be ignored.

5.2 Promotion energy and resonance structures

Whilst hybrid orbitals can be made from atomic orbitals by the methods described above it should be noted that the electron occupancy of these orbitals most favourable to bond formation does not usually correspond to the electronic ground state of the original atom. In carbon for example if the four valence electrons are placed singly in sp^3 hybrid orbitals then this is equivalent to hybridizing an atom in which electrons have been placed, one in each of the $2s$ and $2p$ orbitals: $1s^2 2s^1 2p^3$. Thus the use of hybridized atoms will, in general, involve the implicit use of atoms in electronically excited states. The picture is further complicated by the possibility of many different spectroscopic states arising from the same configuration. In a molecule it is not possible to associate any group of electrons, of certain spin character, with any original atomic arrangement, so that the 'valence state' of an atom, prepared for chemical bonding, will be a suitably weighted average of the spectroscopic states derived from the required electronic configuration. This problem need not detain us further here. Let us merely note the importance of electronically excited states in the use of hybridized atoms. The energy required to go from the ground state to the valence state is called the *promotion energy*. To a first approximation it can be equated with the energy required to go from the ground state configuration to the required excited configuration. Thus to make sp^3 hybrids for carbon a $2s$ electron must be promoted to a $2p$ orbital, i.e. $1s^22s^22p^2 \rightarrow 1s^22s^12p^3$.

In this case the energy required is about 4–5 eV. The promotion energy must be subtracted from the energy of bonds formed using hybrids to give the energy of formation of the molecule from ground state atoms. The valence bond picture of a molecule can be made more accurate by considering other possible structures. For example in methane ionic structures such as $CH_3^+ H^-$ and $H^+ CH_3^-$, as well as nonbonded structures in which no electron promotion has taken place, $H \cdot :CH_2 H \cdot$, could be included. The importance of these structures will depend upon their energy of formation. If it is comparable with the first (sp^3) structure then its contribution to the final picture of the molecule will be quite large. If it is much less stable the contribution will be very small (cf. benzene, p. 52). The inclusion of ionic structures makes up for their neglect in the simple valence bond picture.

If promotion energies are large then the importance of structures based on the hybrids so produced will be diminished since other structures in which no promotion of electrons has been invoked will have comparable energies (even though they may include nonbonded atoms). This becomes more important on the right-hand side of the periodic table where the valence-shell s-orbital becomes tightly bound (Figure 6, p. 26). In the final molecular wave function the contribution of the s-orbitals will be much less than that anticipated by postulating a simple hybrid, for example sp^3 or sp^3d^2. Exactly the same conclusion is reached by the m.o. method since orbitals widely separated in energy (in this case ligand and s-orbitals) will not interact efficiently (p. 54). A corollary is, of course, that valence bond structures that involve atoms in which atomic orbitals, widely different in energy, have been hybridized together will not be a good approximation to a true representation of the molecule. Thus to speak of unique forms of hybridization for elements in groups VI, VII and VIII (e.g. d^2sp^3 for S or Xe in SF_6, XeF_6 etc.) is quite misleading, even though such hybrids might correspond to the correct molecular shape.

5.3 Equivalent orbitals

In the same way as atomic orbitals can be hybridized to yield orbitals localized in a particular region of space so also can molecular orbitals. In a simple stroke this overcomes the conceptual problems associated with delocalized orbitals. To demonstrate the formation of equivalent orbitals let us consider the simple example of two orbitals (ψ_1 and ψ_2) containing a total of four electrons (two of spin α and two of spin β). The over-all wave function can be written as a Slater determinant

$$\Psi = (4!)^{-\frac{1}{2}} \begin{vmatrix} \psi_1\alpha(1) & \psi_1\beta(1) & \psi_2\alpha(1) & \psi_2\beta(1) \\ \psi_1\alpha(2) & \psi_1\beta(2) & \psi_2\alpha(2) & \psi_2\beta(2) \\ \psi_1\alpha(3) & \psi_1\beta(3) & \psi_2\alpha(3) & \psi_2\beta(3) \\ \psi_1\alpha(4) & \psi_1\beta(4) & \psi_2\alpha(4) & \psi_2\beta(4) \end{vmatrix}$$

Now it is a property of determinants that their rows or columns may be added or subtracted from each other without altering the sign or magnitude of the determinant at all. Thus, column 3 can be added to column 1 and column 4 to column 2, to yield (writing down just the top row)

$$\Psi = (4!)^{-\frac{1}{2}} \begin{vmatrix} (\psi_1 + \psi_2)\alpha(1) & (\psi_1 + \psi_2)\beta(1) & \psi_2\alpha(1) & \psi_2\beta(1) \\ \cdots & \cdots & \cdots & \cdots \end{vmatrix}.$$

Take a factor of 2 from each of the first two columns,

$$\Psi = 4(4!)^{-\frac{1}{2}} \begin{vmatrix} \tfrac{1}{2}(\psi_1 + \psi_2)\alpha(1) & \tfrac{1}{2}(\psi_1 + \psi_2)\beta(1) & \psi_2\alpha(1) & \psi_2\beta(1) \\ \cdots & \cdots & \cdots & \cdots \end{vmatrix}.$$

Subtract column 1 from column 3 and column 2 from column 4,

$$\Psi = 4(4!)^{-\frac{1}{2}} \begin{vmatrix} \tfrac{1}{2}(\psi_1 + \psi_2)\alpha(1) & \tfrac{1}{2}(\psi_1 + \psi_2)\beta(1) & \tfrac{1}{2}(\psi_2 - \psi_1)\alpha(1) & \tfrac{1}{2}(\psi_2 - \psi_1)\beta(1) \\ \cdots & \cdots & \cdots & \cdots \end{vmatrix}.$$

Finally, take the factor 4 back into the determinant,

$$\Psi = (4!)^{-\frac{1}{2}} \begin{vmatrix} \sqrt{\tfrac{1}{2}}(\psi_1 + \psi_2)\alpha(1) & \sqrt{\tfrac{1}{2}}(\psi_1 + \psi_2)\beta(1) & \sqrt{\tfrac{1}{2}}(\psi_2 - \psi_1)\alpha(1) & \sqrt{\tfrac{1}{2}}(\psi_2 - \psi_1)\beta(1) \\ \cdots & \cdots & \cdots & \cdots \end{vmatrix}.$$

This is just the Slater determinant that one would write for two electrons of opposite spin in the orbital $(\psi_1 + \psi_2)/\sqrt{2}$ and two electrons of opposite spin in the orbital $(\psi_2 - \psi_1)/\sqrt{2}$. Thus this description is exactly equivalent to the one in which the electron pairs were in orbitals ψ_1 and ψ_2. It may be concluded that linear combinations of orbitals give descriptions of molecules that are just as valid as those using the original orbitals.

The coefficients required for each 'localized' m.o. are found in exactly the same way as for atomic orbitals. Molecular orbitals can be directly substituted for atomic orbitals of the same symmetry, thus in CH_4 the bonding a_1 orbital replaces the s-orbital and the t_2 orbital the p_x, p_y and p_z orbitals in the 'sp^3 hybrid' equations given above.

This ease of interconversion emphasizes the fundamental fact that neither picture of the molecule (localized or delocalized) is more correct than the other. The localized picture emphasizes the presence, at any instant, of two electrons in each chemical bond and further it permits the shape of molecules to be easily explained. Gillespie and Nyholm ('Inorganic stereochemistry', *Quarterly Reviews*, vol. 11 (1957), no. 4, pp. 339–80) have shown that almost all molecular shapes are explicable if it is assumed that electron pairs repel each other in the following order: lone pair–lone pair > lone pair–bonding pair > bonding pair–bonding pair. Because the localized bond picture is built up of numerous pairs of electrons, it is less useful in considering situations where one pair of electrons is broken up, as in electronic excitation or ionization (or indeed radicals and radical ions). In methane there would be four equivalent bonds from which an electron could come to form CH_4^+. Only by invoking the cumbersome notion of resonance between four structures, each with the single electron between carbon and a different hydrogen, is it possible to extend the localized bond picture in this example.

The delocalized model can describe electron excitations very easily since all the orbitals of CH_4 spread over all the 'bonds' in the molecule. But the total amount of charge between each pair of nuclei can only be found after cumbersome calculations. It is not obvious that a pair of electrons is to be found where there is a chemical bond, yet this is so. An advantage of this model is that it emphasizes the actual mobility of electrons in molecules. Although each bond has two electrons, at any one time these will not always be the same two and the delocalized picture shows quite simply how this mobility can take place. In the delocalized picture it is also possible to discuss the energies of the orbitals since they are derived from simple atomic functions and are not hybrids, either of atomic or molecular orbitals.

The two models are therefore quite complementary, depending more on the questions being asked about the molecule, rather than on the molecule itself.

5.3.1 Ethylene

Ethylene has symmetry D_{2h}. A variety of different models may be constructed using different hybridizations at the carbon atom.

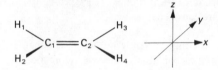

Figure 52   The ethylene molecule

Molecular orbital model. The hydrogen and carbon atoms are numbered as in Figure 52. The z-axis is perpendicular to the plane of the molecule and the C–C axis lies along the x-axis. Carbon orbitals are set parallel to these axes. Then the atomic orbitals belong to the following representations.

Table 17

	E	$C_2(z)$	$C_2(y)$	$C_2(x)$	i	$\sigma_h(xy)$	$\sigma_h(xz)$	$\sigma_h(yz)$	
$\Gamma(H_1, H_2, H_3, H_4)$	4	0	0	0	0	4	0	0	$a_{1g}+b_{1g}+b_{2u}+b_{3u}$
$\Gamma(s_1, s_2)$	2	0	0	2	0	2	2	0	$a_{1g}+b_{3u}$
$\Gamma(p_x(1), p_x(2))$	2	0	0	2	0	2	2	0	$a_{1g}+b_{3u}$
$\Gamma(p_y(1), p_y(2))$	2	0	0	-2	0	2	-2	0	$b_{1g}+b_{2u}$
$\Gamma(p_z(1), p_z(2))$	2	0	0	-2	0	-2	2	0	$b_{1u}+b_{2g}$

Clearly the bonding between these twelve orbitals can be broken down into many smaller groups. The orbitals designated a_{1g} and b_{3u} have axial symmetry about the x-axis. By making simple assumptions about values of Coulomb and resonance integrals it can be shown that two a_{1g} orbitals are bonding and so is one of the b_{3u} orbitals. Of the a_{1g} orbitals one is generally bonding in both C–C and C–H regions whilst the other has a node in all the C–H positions. Conversely the b_{3u} orbital has a node between the carbon atoms. The sum of these effects will be comparable bonding at both C–C and C–H sites. The b_{1g} and b_{2u} orbitals have a plane of symmetry perpendicular to the molecule and containing the x-axis. The xz plane is not perpendicular to any bond and so does not reduce the bonding between atoms. The bonding b_{1g} orbital has no other node but the b_{2u} has a node in the zy plane. The situation is very similar to the π-bonding in buta-1,3-diene and the result is the same, strong bonding at the ends of the molecule between C_1, H_1 and H_2, and also between C_2, H_3 and H_4, but very weak bonding at the middle between C_1 and C_2. The other two orbitals, b_{1u} and b_{2g} are quite independent, having a node in the xy plane, the plane of the molecule. Thus they have π-symmetry. The orbital b_{1u} is bonding and b_{2g} antibonding.

By taking linear combinations of the $2a_{1y} + b_{1g} + b_{2g} + b_{3u}$ bonding orbitals it is possible to generate five new orbitals localized in the regions C_1–H_1, C_1–H_2, C_1–C_2, C_2–H_3 and C_2–H_4. These are the regions of conventional C–C and C–H bonds. If the b_{1u} orbital is also included 'bent' bonds are formed between the carbon atoms. These two models are considered below.

Semi-localized bond model. As a basis for this, currently the most fashionable model for ethylene, we use sp^2 hybridized carbon atoms. This will automatically yield localized electron pair bonds between all the atoms in the xy plane. The sp^2 orbitals from the carbon atoms pointing towards each other, belong to representations a_{1g} and b_{3u}; the former being the bonding C–C σ-orbital. The carbon b_{1u} and b_{2g} orbitals are as before; b_{1u} is the bonding π-orbital. However, the electrons in the π-bond can hardly be said to be localized, concentrated as they are not in one region of space but two; above and below the plane of the molecule. This model of the double bond is useful because it emphasizes the easy polarizability of *one* pair of electrons in the bond. It also suggests that one pair of electrons should be more loosely bound than the other, yielding a lower ionization potential than C–C single bonds, and capable of absorbing ultraviolet light at longer wavelengths. The actual electron distribution in space in the double bond does not, however, reflect this clear-cut distinction between σ- and π-bonds. On the contrary Daudel has shown that there is considerable overlap between σ- and π-orbitals. This aspect may be emphasized if equivalent orbitals are constructed (Figure 53):

$$\psi_1 = \frac{1}{\sqrt{2}}(\sigma + \pi),$$

$$\psi_2 = \frac{1}{\sqrt{2}}(\sigma - \pi).$$

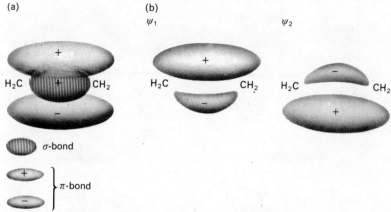

(a)   (b)

σ-bond

π-bond

Figure 53   (a) The $\sigma - \pi$ picture and (b) the equivalent-orbital picture of the bonding in ethylene.

Localized bond model. This corresponds to the original model for ethylene proposed by Bayer utilizing his 'tetrahedral carbon' with C–C contact along one edge. In modern jargon the same model can be constructed using sp^3-hybridized carbon atoms. All the bonds between all atoms are now localized pairs, between the carbon atoms are two 'bent' bonds corresponding to ψ_1 and ψ_2 above.

Since each of these three models for ethylene can be converted one into another by simply taking linear combinations of selected orbitals, they are all equivalent to each other. Whichever is used will depend upon which aspect of the ethylene molecule it is desired to emphasize.

Let us now look briefly at two further examples of orbital equivalence.

5.3.2   *Cyclopropane*

Using sp^3 hybrids the C–C bonds of this molecule are 'bent' suggesting that both a certain strain and unsaturated character might be found in them. Considering just these C–C bonds under the symmetry operations of the relevant point group, D_{3h}, we find they belong to representations a_1' and e'. The three localized C–C bonds are therefore equivalent to three delocalized bonds as shown in Figure 54.

$$\psi_1 = \frac{1}{\sqrt{3}}(C_{12}+C_{23}+C_{31}) \quad a_1'$$

$$\left.\begin{array}{l}\psi_2 = \dfrac{1}{\sqrt{2}}(C_{12}-C_{23}) \\[2mm] \psi_3 = \dfrac{1}{\sqrt{6}}(C_{12}+C_{23}-2C_{31})\end{array}\right\} e'$$

Clearly any unsaturated group attached to C_2 will be able to interact with ψ_2 and unsaturated groups bound to C_1 and C_3 might well be able to interact through ψ_3.

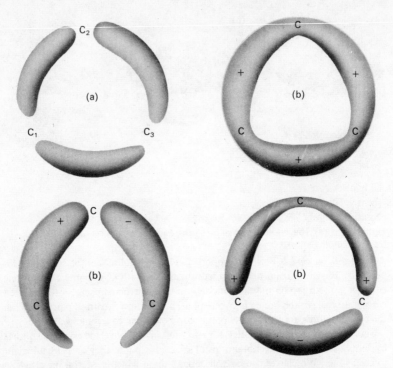

Figure 54　(a) Representation of carbon–carbon bonding in cyclopropane by three bent σ-bonds. (b) Cyclopropane delocalized equivalent orbitals

The partial 'unsaturated' character of the cyclopropane ring is made more apparent by the use of the delocalized equivalent orbitals.

5.3.3　Hyperconjugation

The three C–H bonds of methyl, $-CH_3$ (C_{3v}), can also be regarded as equivalent to three delocalized orbitals of symmetry a_1 and e (Figure 55). Again the possi-

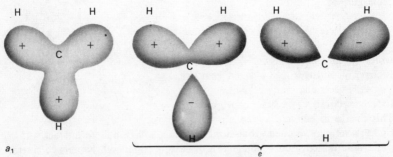

Figure 55　Delocalized equivalent-orbital representation of the methyl group

bility of interaction with an adjacent unsaturated group using one or both of the
e-orbitals can easily be seen.

5.4 Chemical applications

The remainder of this chapter will be devoted to a series of discussions of bonding
problems, in which the symmetry properties of the molecule will be used exten-
sively to simplify the working.

5.4.1 Simple hydrides and oxyanions

Tetrahedral hydrides, BH_4^-, CH_4, NH_4^+, etc., *point group* T_d. Let us assume that
these hydrides are regular tetrahedra and also assume that the chemical bonding
in them arises from $1s$ orbitals on the hydrogens interacting with valence-shell
s-orbitals and p-orbitals on the central atom. The symmetry properties of these
orbitals are shown in Table 18 for XH_4.

Table 18

	E	C_2	S_4	σ_d	C_3	I.R.s
$\Gamma X(s)$	1	1	1	1	1	a_1
$\Gamma X(p_x, p_y, p_z)$	3	-1	-1	$+1$	0	t_2
ΓH_4	4	0	0	2	1	$a_1 + t_2$

The corresponding irreducible representations are shown on the right. The eight-
by-eight determinant may therefore be reduced to one 2×2 a_1 determinant and
three identical 2×2 t_2 determinants

a_1 $\qquad\qquad$ t_2

$$\begin{vmatrix} \alpha_s - E & \beta_1 \\ \beta_1 & \alpha_H - E \end{vmatrix} = 0 \qquad \begin{vmatrix} \alpha_p - E & \beta_2 \\ \beta_2 & \alpha_H - E \end{vmatrix} = 0.$$

Quite obviously four bonding and four antibonding orbitals will result. Now for
most elements that form hydrides, XH_4, $\alpha_H \simeq \alpha_p$ (i.e. about 12 eV), so that t_2
interaction will (resonance integral permitting) give rise to three strong bonds.
However, α_s increases rapidly on going to the right in the periodic table so that
whilst α_ss for boron and carbon are approximately equal to α_H, the disparity
is larger for nitrogen ($\simeq 8$ eV) and even larger for oxygen ($\simeq 15$ eV). Thus even
if β_1 is large in all cases the ability of the a_1 (bonding) orbital to actually contribute effectively to the bonding will diminish as the $\alpha_s - \alpha_H$ difference increases.
This change in bonding power is reflected in thermal stabilities, thus BH_4^- is
stable to 400°C, CH_4 to 600°C but NH_4^+ decomposes below 300° and OH_4^{2+} is
not known. The instability of NH_4^+ is to be contrasted with the great stability
of NH_3 which does not decompose below 2000°C. This clearly shows there is a

great distinction between the three-bond system which depends only on the strong p-bonds and the four-bond system where each orbital is only three-quarters p-bond with only a small extra contribution to the bonding from the s-orbital.

Inorganic oxyanions, XO_3^{n-}, XO_4^{n-}. Boron, carbon and nitrogen all form planar anions, XO_3^{n-} of symmetry D_{3h}. Isolated tetrahedral units XO_4^{n-}, characteristic of second and third row elements are conspicuous by their absence. Boron does form complex structures involving trigonal and tetrahedral coordination and carbon can also be tetrahedrally coordinated by oxygen in esters of the type $C(OR)_4$. Thus the absence of XO_4^{n-} anions from the first row is not due to steric reasons. The orbitals that contribute to bonding in XO_3^{n-} are (ignoring the oxygen $2s$ orbitals as being too tightly bound) listed, together with their symmetry properties, in Table 19.

Table 19

	E	σ_h	C_3	S_3	C_2	σ_v	I.R.s
$X(s)$	1	1	1	1	1	1	a_1'
$X(p_x, p_y)$	2	2	-1	-1	0	0	e'
$X(p_z)$	1	-1	1	-1	-1	1	a_2''
I(O)(three p-orbitals in X–O bonds)	3	-3	0	0	1	1	$a_1' + e'$
II(O)(three p-orbitals perpendicular) to the XO_3^{n-} plane)	3	-3	0	0	-1	1	$a_2'' + e''$
III(O) (three other p-orbitals)	3	3	0	0	-1	-1	$a_2' + e'$

Clearly three X–O bonds can be constructed from X s, p_x and p_y orbitals together with oxygen orbitals from group I ($a_1' + e'$). The other orbitals of symmetry e' from group III will probably not interact since they are at right angles to an X–O bond. The oxygen III orbitals will be equivalent to lone pairs, so also will the e'' orbitals from II but the a_2'' orbital from II will interact with X p_z to form bonding and antibonding orbitals of π-symmetry.

This π-bonding would not be possible in a tetrahedral oxyanion since a first row element would be using all its valence-shell orbitals in σ-bonding. It would seem that oxygen needs to form two covalent bonds (either two σ-bonds, or a σ-bond and a π-bond) if it is to be found in a stable molecule.

In the elements of the second main row of the periodic table, there is the possibility of using $3d$ orbitals in chemical bonding if the central atom is surrounded by electron-attracting ligands. These ligands so polarize the $3s$ and $3p$ orbitals, which screen the $3d$ orbitals in the free atom, that the $3d$ orbitals now experience an increase in effective nuclear charge and so become less diffuse and more tightly bound. Oxygen is strongly electronegative and so will be able to confer bonding potential on the $3d$ orbitals in atoms such as silicon, phosphorus, sulphur, etc. Since these orbitals will be empty let us examine their symmetry properties and also those of the p-orbitals on the oxygen, perpendicular to X–O bonds.

Table 20

	E	C_2	S_4	σ_d	C_3	I.R.s
X(s)	1	1	1	1	1	a_1
X(p)	3	-1	-1	1	0	t_2
X(d)	5	-1	1	1	-1	$e+t_2$
IV(O) (four p-orbitals along X–O bonds)	4	0	0	2	1	a_1+t_2
V(O) (eight p-orbitals perp. X–O bonds)	8	-1	0	0	0	$e+t_1+t_2$
Note:						
$d_{z^2}, d_{x^2-y^2}$	2	-1	2	0	0	e
d_{xz}, d_{yz}, d_{xy}	3	0	-1	1	-1	t_2

The characters for d-orbitals can either be found directly, this is most easily done by considering d_{z^2}, $d_{x^2-y^2}$ and d_{xz}, d_{yz}, d_{xy} separately (see section 5.4.2), or by reference to the general formula for spherical harmonic functions given on p. 170 (or of course by consulting the character table at the back of the book, but that is cheating!).

As in the case of the tetrahedral hydrides, σ-bonds can be formed using X s-orbitals and p-orbitals of symmetries a_1+t_2 together with the ligand orbitals IV. Since d-orbitals have different spatial orientations from p-orbitals, the d-orbitals of t_2 symmetry will probably not contribute greatly to σ-bonding. A π-bond has a nodal plane that contains the axis of the corresponding σ-bond. In this sense the interaction between ligand orbitals V and the X d-orbitals generate π-bonds. A consideration of possible overlap suggests that e-interaction will be stronger than t_2; t_1 orbitals on the ligands will be equivalent to lone pairs.

As with the hydrides it is to be expected that σ-bond strengths will diminish on going to the right in the periodic table due to the increasing ineffectiveness of a_1 bonding. The bonding will generally be strengthened by π-bonding, which should therefore become more important for an anion's existence as the number of the group in the periodic table is increased. This is exactly what is observed, the anions SiO_4^{4-}, PO_4^{3-}, SO_4^{2-}, ClO_4^- show an increasing reluctance to polymerize or even to form any covalent bond to oxygen as this would, of course, interfere with π-bonding. There is a spherical nodal surface in $4d$ orbitals and whilst this may not be exactly in a critical region for bonding, its very existence suggests that $4d$ orbitals may be less efficient at π-bonding than $3d$ orbitals. Again this is clearly born out by experiment. GeO_4^{4-} and AsO_4^{3-} are quite similar to their silicon and phosphorus counterparts but as π-bonding becomes more important so do the anions become less stable. Thus SeO_4^{2-} is a strong oxidizing agent (i.e. easily reduced to SeO_3^{2-} where only p-orbitals are needed for bonding) and BrO_4^- is exceptionally difficult to prepare (see Problem 5.2).

5.4.2 Transition metal complexes

Octahedral complexes of transition metals O_h. One of the most successful applications of the combination of molecular orbital and group theories has been the *ligand field theory* of transition metal complexes. Transition metals are characterized by a partially filled nd shell and further electrons in the $(n+1)s$ (usually two electrons) and $(n+1)p$ (usually one electron) shells (n, principal quantum number). The outermost electrons are easily lost so that most transition metals show ionic valencies of two and/or three. The loss of d-electrons may result in the formation of more highly charged cations. However, whatever the valency of the ion, it is usually surrounded by electron-donating molecules or ions – ligands. Ligand field theory concerns itself with an investigation of the interaction between the ligand orbitals and the outer orbitals of the transition metal ion.

A very common situation is for the central cation to be surrounded by six ligands, at the apices of an octahedron. When the ligands are identical the actual shape and structure of the ligand may be ignored and concentration focused solely on the orbital directed at the metal ion. This set of ligand orbitals has symmetry O_h. The O_h group has operations,

$$E \quad 8C_3 \quad 3C_2 \quad 6C_2 \quad 6C_4 \quad i \quad 8S_6 \quad 3\sigma_h \quad 6\sigma_v \quad 6S_4.$$

The representation for the six ligand orbitals is

$$6 \quad 0 \quad 2 \quad 0 \quad 2 \quad 0 \quad 0 \quad 4 \quad 2 \quad 0.$$

which may be reduced to $a_{1g} + t_{1u} + e_g$.

The representations of the s-orbitals and p-orbitals on the central atom are

(s)	1	1	1	1	1	1	1	1	1	1	a_{1g}
$(p_x p_y p_z)$	3	0	-1	-1	$+1$	-3	0	$+1$	$+1$	-1	t_{1u}.

For the d-orbitals consider the following matrices, the order of orbitals used for E, is retained throughout,

$$
\begin{array}{c}
d_z^2 \\
d_{x^2-y^2} \\
d_{xy} \\
d_{xz} \\
d_{yz}
\end{array}
\quad
E
\begin{bmatrix}
1 & 0 & 0 & 0 & 0 \\
0 & 1 & 0 & 0 & 0 \\
0 & 0 & 1 & 0 & 0 \\
0 & 0 & 0 & 1 & 0 \\
0 & 0 & 0 & 0 & 1
\end{bmatrix}
\quad
\begin{array}{c}
C_3 \text{ (axis passes through} \\
+x, +y, +z \text{ octant)}
\end{array}
\begin{bmatrix}
-\dfrac{1}{2} & -\dfrac{\sqrt{3}}{2} & 0 & 0 & 0 \\
+\dfrac{\sqrt{3}}{2} & -\dfrac{1}{2} & 0 & 0 & 0 \\
0 & 0 & 0 & 1 & 0 \\
0 & 0 & 0 & 0 & 1 \\
0 & 0 & 1 & 0 & 0
\end{bmatrix}
$$

C_2 (z-axis)

$$\begin{bmatrix} 1 & 0 & 0 & 0 & 0 \\ 0 & 1 & 0 & 0 & 0 \\ 0 & 0 & 1 & 0 & 0 \\ 0 & 0 & 0 & -1 & 0 \\ 0 & 0 & 0 & 0 & -1 \end{bmatrix}$$

C_2' (axis passes through origin and $+x, +y, 0$)

$$\begin{bmatrix} 1 & 0 & 0 & 0 & 0 \\ 0 & -1 & 0 & 0 & 0 \\ 0 & 0 & 1 & 0 & 0 \\ 0 & 0 & 0 & 0 & -1 \\ 0 & 0 & 0 & -1 & 0 \end{bmatrix}$$

C_4 (z-axis)

$$\begin{bmatrix} 1 & 0 & 0 & 0 & 0 \\ 0 & -1 & 0 & 0 & 0 \\ 0 & 0 & -1 & 0 & 0 \\ 0 & 0 & 0 & 0 & 1 \\ 0 & 0 & 0 & 1 & 0 \end{bmatrix}$$

i

$$\begin{bmatrix} 1 & 0 & 0 & 0 & 0 \\ 0 & 1 & 0 & 0 & 0 \\ 0 & 0 & 1 & 0 & 0 \\ 0 & 0 & 0 & 1 & 0 \\ 0 & 0 & 0 & 0 & 1 \end{bmatrix}$$

S_6 (as C_3)

$$\begin{bmatrix} -\dfrac{1}{2} & -\dfrac{\sqrt{3}}{2} & 0 & 0 & 0 \\ +\dfrac{\sqrt{3}}{2} & -\dfrac{1}{2} & 0 & 0 & 0 \\ 0 & 0 & 0 & 1 & 0 \\ 0 & 0 & 0 & 0 & 1 \\ 0 & 0 & 1 & 0 & 0 \end{bmatrix}$$

σ_h (perpendicular to the z-axis)

$$\begin{bmatrix} 1 & 0 & 0 & 0 & 0 \\ 0 & 1 & 0 & 0 & 0 \\ 0 & 0 & 1 & 0 & 0 \\ 0 & 0 & 0 & -1 & 0 \\ 0 & 0 & 0 & 0 & -1 \end{bmatrix}$$

$\sigma_v(x = y, z\text{-axis})$

$$\begin{bmatrix} 1 & 0 & 0 & 0 & 0 \\ 0 & -1 & 0 & 0 & 0 \\ 0 & 0 & 1 & 0 & 0 \\ 0 & 0 & 0 & 0 & 1 \\ 0 & 0 & 0 & 1 & 0 \end{bmatrix}$$

S_4 (z-axis)

$$\begin{bmatrix} 1 & 0 & 0 & 0 & 0 \\ 0 & -1 & 0 & 0 & 0 \\ 0 & 0 & -1 & 0 & 0 \\ 0 & 0 & 0 & 0 & -1 \\ 0 & 0 & 0 & -1 & 0 \end{bmatrix}$$

(It is assumed that d_{xy}, d_{xz}, d_{yz} have positive lobes at the sides of the $+x, +y, +z$ octant.) Reduction of these matrices shows that d_{z^2} and $d_{x^2-y^2}$ belong to representation e_g and d_{xy}, d_{xz}, d_{yz} belong to t_{2g}.

The ligand orbitals will therefore interact with s, p and d_{z^2}, $d_{x^2-y^2}$ orbitals but not with d_{xy}, d_{xz}, d_{yz}. If it is assumed that the ligand orbitals are more tightly

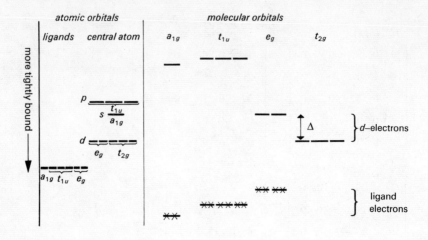

Figure 56 Energy-level diagram for the orbitals of an octahedral complex of a transition metal (Δ is the 'splitting' of the d-orbitals due to the interaction of the ligands).

bound than the nd orbitals then the approximate energy level diagram for the molecular orbitals shown in Figure 56 may be composed.

The most bonding orbitals (a_{1g}, t_{1u}, e_g) may be regarded as filled with ligand electrons. Other electrons would have to be placed in the t_{2g} (non-bonding) and e_g^*, a_{1u}^*, t_{1u}^* orbitals (*, antibonding). Since, in the most stable configuration, electrons will be housed in the most tightly bound orbitals that are available it is reasonable to suppose that the t_{2g} orbitals will be filled first, then e_g^* etc. However, if the interaction between ligand and transition metal orbitals is weak then the energy separation between e_g^* and t_{2g} will be small. If three electrons are disposed, one in each of the t_{2g} orbitals and it is necessary to accommodate more electrons in the complex ion then it might be more economical of energy to put them in the slightly less bonding e_g^* orbitals rather than have them face the electron repulsion that would arise from pairing in the t_{2g} orbitals. In fact complexes of both types are known, 'weak-field' or 'high-spin' complexes in which electrons enter t_{2g} and e_g^* orbitals singly when possible and 'strong-field' or 'low-spin' complexes where the t_{2g}, e_g energy separation is large. In the latter case, electron repulsion notwithstanding, the most stable configurations are achieved by first filling the t_{2g} orbitals completely before putting any electrons in the e_g^* orbitals.

There is an especial stability associated with filled or half-filled shells or spherically symmetrical sub-shells of orbitals. The following electron configurations are therefore favoured: d^0, d^3, d^8, d^{10} and also for a weak field d^5 and for a strong field d^6.

Whether a ligand will interact weakly or strongly with a transition metal ion depends in part on (a) the ligand and (b) the formal charge on the transition metal ion.

(a) Ligands may be thought of as having one particular orbital directed towards the transition metal. If this orbital is tightly bound, weak interaction is to be expected; if loosely bound, strong interaction. In general the strength of a ligand field increases in the series:

halogens (e.g. F^-) < oxygen ligands (e.g. H_2O) < nitrogen ligands (e.g. NH_3) < carbon ligands (e.g. CN^-).

(b) An increase in formal ionic charge will cause the ligands to be more attracted to the central atom. Interaction will therefore increase and ligands which might generate a weak field about a dipositive cation (e.g. $Co(NH_3)_6^{2+}$) could produce a strong field about the same atom if it were oxidized to the III valent state (e.g. $Co(NH_3)_6^{3+}$).

π-bonding in transition metal complexes. So far only the interaction between one orbital from each ligand and the central transition metal has been considered. However, ligands will often also have *p*-orbitals or *d*-orbitals that have nodal planes containing the ligand–cation axis. The overlap of these orbitals with the transition metal orbitals will give rise to π-bonds.

Consider first of all single-atom ligands, each carrying two *p*-orbitals. The representation for the twelve orbitals is

$$12 \quad 0 \quad -4 \quad 0 \quad 0 \quad 0 \quad 0 \quad 0 \quad 0 \quad 0.$$

Thus $\Gamma(12p) = t_{1g} + t_{2g} + t_{1u} + t_{2u}.$[†]

There are no symmetry partners on the central atom for t_{2u} and t_{1g}, which will remain entirely on the ligand atoms. Being derived from *p*-orbitals on the ligands that are perpendicular to the Cartesian axes, t_{1u} will not interact with the t_{1u} orbitals of the central atom as strongly as the ligand orbitals ($t_{1u}\sigma$) directed at the transition metal cation. But t_{2g} orbitals on the central atom are directed towards the side of each ligand – just where the *p*-orbitals now being considered are to be found. Strong interaction may therefore be anticipated. The over-all effect of π-bonding from the twelve *p*-orbitals is to generate bonding and anti-bonding t_{2g} orbitals and nonbonding t_{1g}, t_{1u} and t_{1g} orbitals. If the *p*-orbitals are filled on the ligand atoms and the transition metal is d^0 then π-bonding will help to stabilize the complex. The lower the *d*-electron population (i.e. the higher the valency) on the central atom the more effective will this π-bonding be. Ligands such as oxygen and fluorine do, in fact, often invoke the highest possible formal valency of an atom and π-bonding probably stabilizes their complexes. Highly charged cations are, of course, very small and steric factors probably constrain the oxygen complexes of the first-row transition metals to adopt tetrahedral coordination.

e.g. $VO_4^{3-}(d^0)$ $CrO_4^{2-}(d^0)$ $MnO_4^-(d^0)$ $MnO_4^{2-}(d^1)$ $FeO_4^{2-}(d^2)$.

[†] Wave functions corresponding to these representations are given on p.147 in the discussion of the B_6^{2-} complex.

If the p-orbitals on the ligands were vacant then interaction of this type would provide stable bonding orbitals for these electrons normally found in the t_{2g} nonbonding orbitals. In this way electron-rich transition metal atoms (i.e. low valency states) could be stabilized by complex formation. Rather than vacant p-orbitals on ligand atoms, vacant molecular orbitals in more complex ligands and vacant d-orbitals on atoms such as phosphorus are more common. For the ligands CO or CN^- the bonding molecular orbitals will belong to representations $t_{1g}, t_{2g}, t_{1u}, t_{2u}$; but these orbitals will be tightly bound and unlikely to interact with d-orbitals on the central atoms. Each ligand also carries a pair of anti-bonding (vacant) molecular orbitals and these orbitals from all six ligands will belong to the same symmetry representations as the bonding molecular orbitals. The antibonding orbitals will have energies somewhat comparable with d-orbitals. This interaction will generate new t_{2g} molecular orbitals. One set will be bonding relative to the transition-metal d-orbitals so that up to six d-electrons could be accommodated in these bonding π-orbitals. Ligands are normally regarded as donating electrons to the transition metal. But in this type of π-bonding the direction of donation is reversed so it is often called 'back bonding'. Zero-valent chromium and molybdenum can both be regarded as d^6 and both form hexa-carbonyls – $Cr(CO)_6$ and $Mo(CO)_6$ – in which back bonding is undoubtedly important. Many transition metals are found in subnormal valency states when surrounded by phosphorus ligands, e.g. PCl_3, $P(C_6H_5)_3$. Such complexes probably owe their existence to the stabilizing effect of π-bonds involving vacant $3d$ orbitals on the phosphorus.

5.5 π-Bonding in organic molecules

A large number of organic molecules are known in which all the carbon atoms lie more or less in a plane. Bonds that are symmetric with respect to reflection in this plane (i.e. $\chi(\sigma) = +1$) are σ-bonds; those that are antisymmetric ($\chi(\sigma) = -1$) are π-bonds, as in ethylene, for example, discussed on page 120. When many double bonds are arranged so that the π-orbitals can interact together, *conjugation* is said to occur. The electrons that move in these orbitals are particularly amenable to simple theoretical calculations since they are subject to a more even potential than σ-electrons.

The symmetry of a molecule is often a great help in simplifying calculations of π molecular orbitals, as will be demonstrated in the following examples.

5.5.1   *Buta-1,3-diene*

The four p-orbitals that interact to give π molecular orbitals have already been discussed in Chapter 4 (p. 104). Since there is no central atom, it is necessary to group together orbitals that are equivalent with respect to the symmetry opera-tions of the molecule. From the *trans* isomer (C_{2h}) it has been shown that $\Gamma(C_1, C_4) = \Gamma(C_2, C_3) = a_u + b_g$. The same pairing of orbitals is applicable to the *cis* isomer (C_{2v}), $\Gamma(C_1, C_4) = \Gamma(C_2, C_3) = a_2 + b_2$. 'Molecular' orbitals can be constructed:

$$\text{I.R. } trans \quad \text{I.R. } cis$$

$$\psi_i = \frac{1}{\sqrt{2}}(C_1 + C_4) \qquad a_u \qquad b_2$$

$$\psi_{ii} = \frac{1}{\sqrt{2}}(C_1 - C_4) \qquad b_g \qquad a_2$$

$$\psi_{iii} = \frac{1}{\sqrt{2}}(C_2 + C_3) \qquad a_u \qquad b_2$$

$$\psi_{iv} = \frac{1}{\sqrt{2}}(C_2 - C_3) \qquad b_g \qquad a_2$$

In Chapter 2 the energies of the π molecular orbitals were found by solving a 4×4 determinant. This procedure may now be simplified. For the a_u or b_2 orbitals,

$$\begin{matrix} \psi_i & \psi_{iii} \end{matrix}$$
$$\begin{vmatrix} \alpha_1 - E & \beta_1 \\ \beta_1 & \alpha_2 - E \end{vmatrix} = 0.$$

$$\alpha_1 = \int \psi_i \mathscr{H} \psi_i \, d\tau = \tfrac{1}{2}(\int C_1 \mathscr{H} C_1 \, d\tau + 2 \int C_1 \mathscr{H} C_4 \, d\tau + \int C_4 \mathscr{H} C_4 \, d\tau)$$

$$= \alpha$$

$$= \text{Coulomb integral for carbon } 2p \text{ orbital.}$$

($\int C_1 \mathscr{H} C_4 \, d\tau = 0$ since orbitals nonadjacent.)

$$\alpha_2 = \int \psi_{iii} \mathscr{H} \psi_{iii} \, d\tau = \tfrac{1}{2}(\int C_2 \mathscr{H} C_2 \, d\tau + 2 \int C_2 \mathscr{H} C_3 \, d\tau + \int C_3 \mathscr{H} C_3 \, d\tau)$$

$$= \alpha + \beta,$$

where β = resonance integral between adjacent p-orbitals.

$$\beta_1 = \int \psi_i \mathscr{H} \psi_{iii} \, d\tau$$

$$= \tfrac{1}{2}(\int C_1 \mathscr{H} C_2 \, d\tau + \int C_1 \mathscr{H} C_3 \, d\tau + \int C_4 \mathscr{H} C_2 \, d\tau + \int C_4 \mathscr{H} C_3 \, d\tau)$$

$$= \beta.$$

If $x = (\alpha - E)/\beta$, then the determinant can be written

$$\begin{vmatrix} x & 1 \\ 1 & x+1 \end{vmatrix} = 0$$

i.e. $x^2 + x - 1 = 0$

and $\qquad x = \tfrac{1}{2}(-1 \pm \sqrt{5})$

$$= -1 \cdot 62, \; +0 \cdot 62.$$

Similarly the b_g or a_2 determinant is

$$\begin{matrix} \psi_{ii} & \psi_{iv} \\ \begin{vmatrix} x & 1 \\ 1 & x-1 \end{vmatrix} & = 0 \end{matrix}$$

i.e. $x^2 - x - 1 = 0$

$$x = \tfrac{1}{2}(1 \pm \sqrt{5})$$
$$= +1\cdot62, \ -0\cdot62.$$

Thus energy levels and charge distributions are the same in both isomers of buta-1,3-diene. However, the differences in symmetry will be important when discussing electronic transitions between the π molecular orbitals (p. 161).

5.5.2   *π-Orbitals in cyclic hydrocarbon molecules and ions C_nH_n*

For convenience it will be assumed that these molecules belong to the point group D_{nh}. This can be true for small values of n, up to say eight. Thereafter, the simple regular polygon is no longer the most stable configuration. Even so, to assume the regular shape will not alter the results, at least to a first approximation.

If n is even, the following operations upon the n p-orbitals will yield non-zero characters:

$$\begin{array}{cccc} E & C_2' & \sigma_h & \sigma_v' \\ n & -2 & -n & +2 \end{array}$$

C_2' is a twofold axis linking carbon atoms on opposite sides of the polygon. This axis and the principal axis are contained by σ_v'. The representation for the n p-orbitals can be reduced to

$$a_{2u} + e_{1g} + e_{2u} + e_{3g} + e_{4u} + \ldots + b_{2g} \text{ or } b_{2u}.$$

The gerade (g) and ungerade (u) representations alternate until the reduction of the original representation, $\chi(E) = n$, is complete.

When n is odd, the non-zero characters in the representation of the p-orbitals are

$$\begin{array}{cccc} E & C_2' & \sigma_h & \sigma_v \\ n & -1 & n & +1 \end{array}$$

which reduces to irreducible representations $a_2'' + e_1'' + e_2'' + \ldots$.

The coefficients of the degenerate orbitals (*not* normalized) are shown in Figure 57; δ is related to the 'angle of orthogaonality' (p. 92) and is the subscript number of the degenerate representation (i.e. for e_2'' or e_{2u}, $\delta = 2$). θ is related to n simply by $\theta° = (360/n)°$. The m.o.s are $\psi_1 = \sum_n \phi_k \sin k\delta\theta$ and $\psi_2 = \sum_n \phi_k \cos k\delta\theta$ where ϕ_k is the carbon $2p$ orbital at atom k (see also problem 1.2).

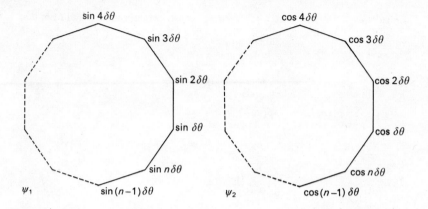

Figure 57   Representation of the coefficients (not normalized) of the degenerate orbitals in cyclic hydrocarbons $C_n H_n$

To check that orbitals with these coefficients transform in the correct way, let us investigate the rotation C_n. In ψ_1 any general coefficient, $\sin k\delta\theta$ is moved to the site of coefficient $\sin(k+1)\delta\theta$.

Thus $C_n(\sin k\delta\theta) = \sin(k+1)\delta\theta$,

i.e. $C_n(\sin k\delta\theta) = \cos \delta\theta \sin k\delta\theta + \sin \delta\theta \cos k\delta\theta$.

These equations may be summarized in matrix form

$$C_n\begin{bmatrix} \sin k\delta\theta \\ \cos k\delta\theta \end{bmatrix} = \begin{bmatrix} \cos \delta\theta & \sin \delta\theta \\ -\sin \delta\theta & \cos \delta\theta \end{bmatrix} \begin{bmatrix} \sin k\delta\theta \\ \cos k\delta\theta \end{bmatrix}.$$

Since this equation has been developed for any coefficient in ψ_1 and ψ_2 it follows that ψ_1 and ψ_2 will transform in the required way.

The functions may be normalized, by multiplying by the following factors:

for ψ_1 $\left[\sum_{k=1}^{n} \sin^2 k\delta\theta \right]^{-1}$

and for ψ_2 $\left[\sum_{k=1}^{n} \cos^2 k\delta\theta \right]^{-1}$.

It can also be shown by summing the products of the coefficients at each atom from ψ_1 and ψ_2 that $\int \psi_1 \psi_2 \, d\tau = 0$: the two orbitals are orthogonal.

It is interesting to note that when $\delta = 0$ or n all the coefficients of ψ_1 are zero but those of ψ_2 are all $+1$, corresponding to the a_{2u} or a_2'' orbital. When n is even $\delta = (\frac{1}{2}n)$ also yields one orbital, again from ψ_2: the un-normalized coefficients in $b_{2(u \text{ or } g)}$ alternate $+1, -1, +1$, etc., from atom to atom around the ring.

Calculation of π-orbital energies. The energy of any orbital can be found from the equation $E = \dfrac{\int \psi \, \mathscr{H} \psi \, d\tau}{\int \psi \psi \, d\tau}$. Substituting ψ_2,

$$E = \frac{\int \sum\limits_{k=1}^{n} \phi_k \cos k\delta\theta \, \mathscr{H}\left(\sum\limits_{k=1}^{n} \phi_k \cos k\delta\theta\right) d\tau}{\int \sum\limits_{k=1}^{n} \phi_k \cos k\delta\theta \sum\limits_{k=1}^{n} \phi_k \cos k\delta\theta \, d\tau}$$

$$E = \frac{\sum\limits_{k=1}^{n} \left(\int \phi_k \mathscr{H} \phi_k \, d\tau\right) \cos^2 k\delta\theta + 2 \sum\limits_{k=1}^{n} \left(\int \phi_k \mathscr{H} \phi_{k+1} \, d\tau\right) \cos k\delta\theta \cos(k+1)\delta\theta}{\sum\limits_{k=1}^{n} \left(\int \phi_k \phi_k \, d\tau\right) \cos^2 k\delta\theta}.$$

Resonance integrals between nonadjacent atoms have been ignored and the neglect of overlap approximation used.

$\alpha = \int \phi_k \mathscr{H} \phi_k \, d\tau = $ Coulomb integral for carbon $2p$ orbitals,

$\beta = \int \phi_k \mathscr{H} \phi_{k+1} \, d\tau = $ resonance integral between adjacent p-orbitals.

Now $\quad \cos k\delta\theta \cos(k+1)\delta\theta = \cos^2 k\delta\theta \cos \delta\theta - \sin \delta\theta \sin k\delta\theta \cos k\delta\theta$

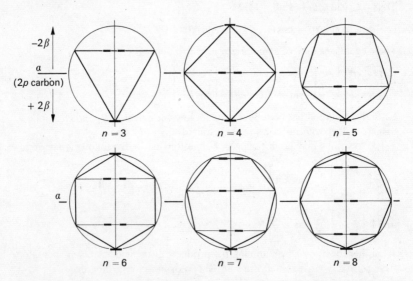

Figure 58 Diagrams representing the energy levels of π molecular orbitals in cyclic hydrocarbon species $C_n H_n$ for values of n from 3 to 8. The level of the horizontal diameter represents the value of the Coulomb integral for the carbon $2p$ orbitals

$$E = \frac{\alpha \sum_{k=1}^{n} \cos^2 k\delta\theta + 2\beta \cos \delta\theta \sum_{k=1}^{n} \cos^2 k\delta\theta - \sin \delta\theta \frac{1}{2} \sum_{k=1}^{n} \sin 2k\delta\theta}{\sum_{k=1}^{n} \cos^2 k\delta\theta}.$$

But $\sum_{k=1}^{n} \sin 2k\delta\theta = 0$ since $\theta^\circ = \dfrac{360^\circ}{n}$,

thus $E = \alpha + 2\beta \cos \delta\theta = \alpha + 2\beta \cos \dfrac{360\delta^\circ}{n}$. **5.3**

The values of δ from 1 to n will give the n energy levels of the system. Because of the form of equation **5.3**, the molecular orbital energy levels of cyclic polyenes may be very simply arrayed by enscribing a circle about the polygon with the apex downwards and dropping perpendiculars to the *vertical* axis from each vertex. The horizontal diameter represents the α of the $2p$ carbon orbitals. Intercepts to the vertical axis below this line give the energies of bonding molecular orbitals in units of 2β; intercepts to the vertical axis above this line yield energies of antibonding orbitals in units of -2β (Figure 58).

$C_5H_5^-$, *Cyclopentadienyl anion.* Using these general methods the energies and equations for the five π molecular orbitals are (Figure 59):

$\psi(a_2'') = \dfrac{\phi_1 + \phi_2 + \phi_3 + \phi_4 + \phi_5}{\sqrt{5}}$ $\qquad E = \alpha + 2\beta$

$\left.\begin{aligned} \psi(e_1'')_i &= \sqrt{(\tfrac{2}{5})}(0\cdot309\phi_1 - 0\cdot809\phi_2 - 0\cdot809\phi_3 + 0\cdot309\phi_4 + \phi_5) \\ \psi(e_1'')_{ii} &= \sqrt{(\tfrac{2}{5})}(0\cdot951\phi_1 + 0\cdot588\phi_2 - 0\cdot588\phi_3 - 0\cdot951\phi_4) \end{aligned}\right\} \quad E = \alpha + 0\cdot618\beta$

$\left.\begin{aligned} \psi(e_2'')_i &= \sqrt{(\tfrac{2}{5})}(-0\cdot809\phi_1 + 0\cdot309\phi_2 + 0\cdot309\phi_3 - 0\cdot809\phi_4 + \phi_5) \\ \psi(e_2'')_{ii} &= \sqrt{(\tfrac{2}{5})}(+0\cdot588\phi_1 - 0\cdot951\phi_2 + 0\cdot951\phi_3 - 0\cdot588\phi_4) \end{aligned}\right\} \quad E = \alpha - 1\cdot618\beta$

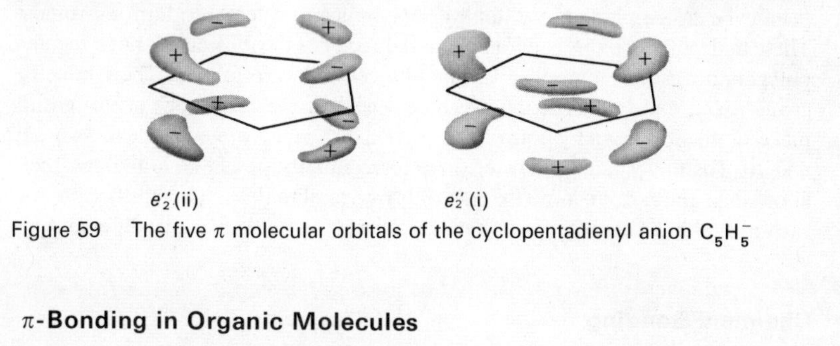

Figure 59   The five π molecular orbitals of the cyclopentadienyl anion $C_5H_5^-$

C_6H_6, *Benzene.* The π molecular orbitals and their energies are (Figure 60):

$$\psi(a_{2u}) = \frac{\phi_1 + \phi_2 + \phi_3 + \phi_4 + \phi_5 + \phi_6}{\sqrt{6}} \qquad E = \alpha + 2\beta,$$

$$\psi(e_{1g})_i = \frac{\phi_1 - \phi_2 - 2\phi_3 - \phi_4 + \phi_5 + 2\phi_6}{\sqrt{12}}$$

$$\left.\begin{array}{l} \\ \\ \end{array}\right\} E = \alpha + \beta,$$

$$\psi(e_{1g})_{ii} = \frac{\phi_1 + \phi_2 - \phi_4 - \phi_5}{2}$$

$$\psi(e_{2u})_i = \frac{-\phi_1 - \phi_2 + 2\phi_3 - \phi_4 - \phi_5 + 2\phi_6}{\sqrt{12}}$$

$$\left.\begin{array}{l} \\ \\ \end{array}\right\} E = \alpha - \beta,$$

$$\psi(e_{2u})_{ii} = \frac{\phi_1 - \phi_2 + \phi_4 - \phi_5}{2}$$

$$\psi(b_{2g}) = \frac{\phi_1 - \phi_2 + \phi_3 - \phi_4 + \phi_5 - \phi_6}{\sqrt{6}} \qquad E = \alpha - 2\beta.$$

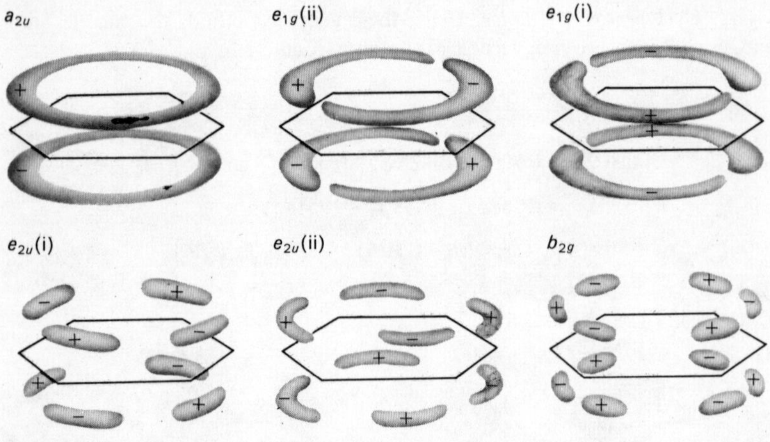

Figure 60   The π molecular orbitals of benzene

$C_{10}H_8$, *Naphthalene.* The ten $2p$ carbon orbitals (numbered as in Figure 61) perpendicular to the plane of the naphthalene molecule interact to form π-orbitals. These molecular orbitals cannot be found directly as the molecule is not a regular polygon, nor is the simplifying feature of a central atom present. The symmetry group is $\boldsymbol{D_{2h}}$ and by inspection it can be seen that the operations of this group make atoms 2, 3, 6 and 7 equivalent. Also 1, 4, 5 and 8 are equivalent, as are 9 and 10. The molecular orbitals of these three subgroups can be found and then allowed to interact to form the π-orbitals of naphthalene. The coefficients for each orbital in a subgroup must have the same magnitude but they differ in sign.

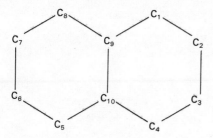

Figure 61   Carbon skeleton of naphthalene

Table 21

p-orbitals	E	$C_2(z)$	$C_2'(y)$	$C_2''(x)$	i	σ_h	$\sigma_v(y)$	$\sigma_v'(x)$	
1, 4, 5, 8	4	0	0	0	0	-4	0	0	$a_{1u}+b_{1u}+b_{2g}+b_{3g}$
2, 3, 6, 7	4	0	0	0	0	-4	0	0	$a_{1u}+b_{1u}+b_{2g}+b_{3g}$
9, 10	2	0	0	-2	0	-2	$+2$	0	$b_{1u}+b_{2g}$

The molecular orbitals are

b_{1u} $\psi_1 = \tfrac{1}{2}(\phi_1 + \phi_4 + \phi_5 + \phi_8)$,
  $\psi_2 = \tfrac{1}{2}(\phi_2 + \phi_3 + \phi_6 + \phi_7)$,
  $\psi_3 = \dfrac{\phi_9 + \phi_{10}}{\sqrt{2}}$,

a_{1u} $\psi_4 = \tfrac{1}{2}(\phi_1 - \phi_4 + \phi_5 - \phi_8)$,
  $\psi_5 = \tfrac{1}{2}(\phi_2 - \phi_3 + \phi_6 - \phi_7)$,

b_{2g} $\psi_6 = \tfrac{1}{2}(\phi_1 - \phi_4 - \phi_5 + \phi_8)$,
  $\psi_7 = \tfrac{1}{2}(\phi_2 - \phi_3 - \phi_6 + \phi_7)$,
  $\psi_8 = \dfrac{\phi_9 - \phi_{10}}{\sqrt{2}}$,

b_{3g} $\psi_9 = \tfrac{1}{2}(\phi_1 + \phi_4 - \phi_5 - \phi_8)$,
  $\psi_{10} = \tfrac{1}{2}(\phi_2 + \phi_3 - \phi_6 - \phi_7)$.

Using the conventional symbolism and making the usual approximations the determinants are

$$b_{1u} \quad \begin{matrix} & \psi_1 & \psi_2 & \psi_3 \\ \psi_1 & x & 1 & 2 \\ \psi_2 & 1 & x+1 & 0 \\ \psi_3 & 2 & 0 & x+1 \end{matrix} = 0,$$

i.e $x(x+1)^2 - (x+1) + 2\{-2(x+1)\} = 0,$

$$(x+1)(x^2+x-3) = 0.$$

Hence $\quad x = -1$ and $\frac{1}{2}(-1\pm\sqrt{13}).$

$$a_{1u} \quad \begin{array}{c} \psi_4 \quad \psi_5 \\ \psi_4 \\ \psi_5 \end{array} \begin{vmatrix} x & 1 \\ 1 & x-1 \end{vmatrix} = 0$$

i.e. $\qquad x^2 - x - 1 = 0.$

Thus $\quad x = \frac{1}{2}(+1\pm\sqrt{5}).$

$$b_{2g} \quad \begin{array}{c} \psi_6 \quad \psi_7 \quad \psi_8 \\ \psi_6 \\ \psi_7 \\ \psi_8 \end{array} \begin{vmatrix} x & 1 & 2 \\ 1 & x-1 & 0 \\ 2 & 0 & x-1 \end{vmatrix} = 0$$

i.e. $\quad x(x-1)^2 - (x-1) + 2\{-2(x-1)\} = 0$

$$(x-1)(x^2-x-3) = 0$$

$x = +1$ and $\frac{1}{2}(+1\pm\sqrt{13})$

$$b_{3g} \quad \begin{array}{c} \psi_9 \quad \psi_{10} \\ \psi_9 \\ \psi_{10} \end{array} \begin{vmatrix} x & 1 \\ 1 & x+1 \end{vmatrix} = 0$$

$$x^2 + x - 1 = 0$$

i.e. $\quad x = \frac{1}{2}(-1\pm\sqrt{5}).$

The energy levels of the π-orbitals of naphthalene may be summarized (in units of β, negative values bonding): $-2\cdot303$ (b_{1u}), $-1\cdot618$ (b_{3g}), $-1\cdot303$ (b_{2g}), -1 (b_{1u}), $-0\cdot636$ (a_{1u}), $+0\cdot636$ (b_{3g}), $+1$ (b_{2g}), $+1\cdot303$ (b_{1u}), $+1\cdot618$ (a_{1u}), $+2\cdot303$ (b_{2g}).

$C_5H_5^- Fe^{2+} C_5H_5^-$, *Ferrocene.* Aromatic ring systems have suitable arrays of bonding and antibonding π molecular orbitals to interact with transition metal ions to form complexes. In ferrocene a ferrous ion is sandwiched between two cyclopentadienyl anions; the rings are staggered so that the molecule belongs to the point group D_{5d}. To describe the bonding between the anions and the central ion the two cyclopentadienyl systems must be considered together. The characters that describe the action of the various D_{5d} operators on the cyclopentadienyl and iron orbitals are given in Table 22.

Table 22

Orbitals	E	$2C_5$	$2C_5^2$	$5C_2'$	i	$2S_{10}$	$2S_{10}^3$	$5\sigma_d$	Representations in D_{5d}
a_2'' on the two anions	2	2	2	0	0	0	0	0	$a_{1g}+a_{2u}$
e_1'' on the two anions	4	$4\cos 72°$	$4\cos 144°$	0	0	0	0	0	$e_{1g}+e_{1u}$
e_2'' on the two anions	4	$4\cos 144°$	$4\cos 72°$	0	0	0	0	0	$e_{2g}+e_{2u}$
Fe, s	1	1	1	1	1	1	1	1	a_{1g}
Fe, ps	3	$1+2\cos 72°$	$1+2\cos 144°$	-1	-3	$-1+2\cos 36°$	$-1+2\cos 108°$	0	$a_{2u}+e_{1u}$
Fe, ds	5	$1+2\cos 72° +2\cos 144°$	$1+2\cos 144° +2\cos 288°$	1	5	$1-2\cos 36° +2\cos 72°$	$1-2\cos 36° +2\cos 216°$	0	$a_{1g}+e_{1g}+e_{2g}$

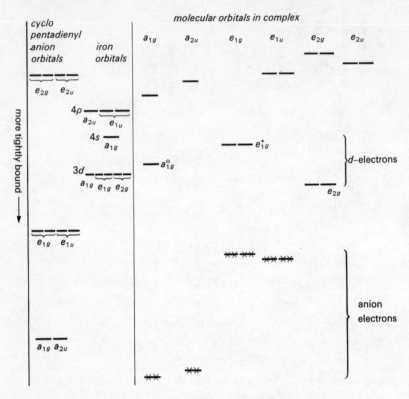

Figure 62   Qualitative energy-level diagram for the molecular orbitals of ferrocene

If it is now assumed that the bonding orbitals in cyclopentadienyl are all more tightly bound than the valency orbitals of iron but that the antibonding orbitals are less tightly bound, then a qualitative energy level diagram may be drawn, Figure 62. This array of bonding and antibonding molecular orbitals may also be used to discuss the bonding between any transition metal and a pair of cyclopentadienyl anions. Orbitals a_{1g}, a_{2u}, e_{1g} and e_{1u} may be regarded as filled with the ligand electrons. The valency electrons that were housed on the transition element must now be disposed amongst the remaining reasonably tightly bound orbitals. Relative to the original d-orbital ionization potential, the a_{1g}^0 orbital will be slightly antibonding and the e_{2g} will be quite appreciably bonding. The e_{1g} orbitals are antibonding and will become more so as the d-orbitals become more tightly bound. The spin character and suggested orbital occupancy of some complexes of this type are shown in Table 23. It is curious that chromium Cr^{2+} does not adopt the e_{2g}^4 configuration. If the splitting between the a_{1g}^0 and e_{2g} levels is small then the compromise $(a_{1g}^0)^2 e_{2g}^2$ probably yields the most stable configuration. The contrast between high-spin Mn^{2+} and low-spin Fe^{2+} is striking. In both cases factors other than simple orbital energy levels are obviously important. For Mn^{2+} the five d-electrons benefit from the stability of a symmetrically half-filled shell of orbitals whilst for Fe^{2+} stability is associated with the completely filled orbitals e_{2g} and a_{1g}.

Table 23

Metal cation	d-electrons	Orbitals occupied	Number of unpaired electrons
V^{3+}	2	e_{2g}^2	2
V^{2+}, Cr^{3+}	3	$e_{2g}^2(a_{1g}^0)^1$	3
Cr^{2+}	4	$e_{2g}^2(a_{1g}^0)^2$	2
Mn^{2+}	5	$e_{2g}^2(a_{1g}^0)^1 e_{1g}^2$	5
Fe^{3+}	5	$e_{2g}^4 a_{1g}^1$	1
Fe^{2+}, Co^{3+}	6	$e_{2g}^4(a_{1g}^0)^2$	0
Co^{2+}	7	$e_{2g}^4(a_{1g}^0)^2 e_{1g}^1$	1
Ni^{2+}	8	$e_{2g}^4(a_{1g}^0)^2 e_{1g}^2$	2

The complexes formed between cyclopentadienyl anions and transition metal ions are not necessarily neutral – a range of charged species is known. Also cyclopentadiene is by no means unique amongst the $C_n H_n$ systems in forming complexes of this type. Related compounds involving $C_4 H_4^{2-}$, $C_6 H_6$ and $C_7 H_7^+$ have been isolated. Finally it is not necessary for these ligands to be bound to the transition metal in symmetric pairs. Many complexes are known which contain only one $C_n H_n$-type ligand together with a variety of other ligands.

5.6 'Electron-deficient' molecules

The elements of groups I, II and III are not endowed by nature with enough electrons to be able to form simple electron-pair bonds utilizing all their valence-shell orbitals. Should boron for example try to form four bonds with four hydrogen atoms BH_4^0 (seven electrons) would result. The anion BH_4^- is however known, isoelectronic with methane. In all molecules that can be described in terms of localized electron-pair bonds the interaction of atomic orbitals generates an equal number of bonding and antibonding orbitals. If a requirement of relative chemical stability is that all bonding orbitals shall be filled, then clearly atoms from groups I, II and III unless they are in combination with electron-rich atoms will form molecules with vacancies in bonding orbitals. A simple way of overcoming this difficulty would be for such atoms to try to form systems in which more antibonding than bonding orbitals were found.

Such systems are, in fact, generated when more than two orbitals all overlap with each other. An example of overlap of this type has already been considered in the π-orbitals of the cyclopropenium cation (p. 38). Three p-orbitals all interact with each other: one bonding and two antibonding orbitals are formed. If electron-deficient atoms are to attempt covalent bonding they will therefore tend to form structures in which each orbital can overlap with as many other orbitals as possible.

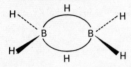

Figure 63   Structure of diborane

5.6.1   *Diborane*
(See also problem 4.4.) The structure is shown in Figure 63. It will simplify the calculations if it is assumed that the boron is sp^2 hybridized. This should not lead to erroneous results since $2s$ and $2p$ orbitals have similar energies (Figure 6, p. 26). The four peripheral hydrogens may be regarded as bound to the boron atoms by conventional bonds. The representations for the other orbitals are shown in Table 24.

Table 24

Orbitals	E	$C_2(z)$	$C_2'(y)$	$C_2''(x)$	i	$\sigma_h(xy)$	$\sigma_h(xz)$	$\sigma_h(yz)$	
two sp^2 hybrids	2	0	2	0	0	2	0	2	$a_{1g}+b_{2u}$
two p-orbitals	2	0	-2	0	0	-2	0	2	$b_{3g}+b_{1u}$
two hydrogen orbitals	2	2	0	0	0	0	2	2	$a_{1g}+b_{1u}$

The two sp^2 hybrid orbitals will interact strongly; the a_{1g} orbital will be strongly bonding and the b_{2u} strongly antibonding. Furthermore, the a_{1g} orbital will overlap with the hydrogen orbital of the same symmetry so that a new, very strongly bonding orbital and an antibonding orbital are formed. Similarly the two boron p-orbitals will form bonding (b_{1u}) and antibonding (b_{3u}) π-bonds. The former will interact with the hydrogen b_{1u} orbital forming a strongly bonding orbital and an antibonding orbital. The over-all picture is the formation of only two bonding orbitals between the four atoms, but these orbitals are quite strongly bonding.

So even though each atom only contributes one electron a stable structure can be formed. An exactly equivalent picture involving localized, bent B–H–B bridge bonds is developed in the next section.

5.6.2 Pentaborane, B_5H_9.

The structure of this pentaborane is shown in Figure 19, p. 61. As in the case of diborane a consideration of the structure is greatly simplified by the use of hybrid orbitals. Three types of bonding may be distinguished in the molecule. Simple B–H bonds, B–H–B linkages, and the bonding of the apical BH unit to the rest of the molecule. This division in no way implies that it is not possible to describe molecular orbitals that extend over the whole molecule, rather that the features of bonding may be more clearly examined if the molecule is dissected in this way.

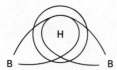

Figure 64   Overlap of three atomic orbitals forming the B–H–B bond

Each B–H–B linkage may be thought of as the overlap of three orbitals (Figure 64). The energy levels of the resulting molecular orbitals may be found by solving:

$$\begin{vmatrix} \alpha_B - E & \beta & \beta' \\ \beta & \alpha_H - E & \beta \\ \beta' & \beta & \alpha_B - E \end{vmatrix} = 0,$$

where α_B = Coulomb integral for boron sp^3 hybrid,
α_H = Coulomb integral for hydrogen $1s$ orbital,
β = resonance integral for boron hybrid–hydrogen interaction
and β' = resonance integral for boron hybrid–boron hybrid interaction.

This is equivalent to

$$\begin{vmatrix} x & 1 & k \\ 1 & x+\varepsilon & 1 \\ k & 1 & x \end{vmatrix} = 0,$$

where $x = \dfrac{\alpha_B - E}{\beta}$, $k = \dfrac{\beta'}{\beta}$

and $\alpha_H = \alpha_B + \varepsilon\beta$; ε and k being constants.

143   'Electron-deficient' Molecules

Expanding the determinant:

$$(x-k)\{(x+k)(x+\varepsilon)-2\} = 0.$$

Thus $\quad\quad\quad\quad\quad\quad x = k$

or $\quad\quad\quad\quad\quad\quad x = \frac{1}{2}[-(k+\varepsilon)\pm\sqrt{\{(k-\varepsilon)^2+8\}}].$

In this example it seems reasonable to assume that both k and ε will be small (i.e. about one or less) so that only one bonding orbital will be formed. This calculation is analogous to that for the cyclopropenium cation. Thus two electrons are required for the bonding orbital, a bonding orbital that will link together three atoms. Diborane may be regarded as being linked by two such three-centre bonds – a description equivalent to that given above.

The bonding of the apical boron may be simplified by considering the representations of the orbitals involved.

Table 25

Orbitals	E	C_2	$2C_4$	$2\sigma_v$	$2\sigma_v'$	
four sp^3 hybrids – one from each basal B atom	4	0	0	2	0	a_1+e+b_1
apical boron sp hybrid – down z-axis	1	1	1	1	1	a_1
two p-orbitals	2	-2	0	0	0	e

If ϕ_1, ϕ_2, ϕ_3 and ϕ_4 represent the four sp^3 hybrids, one from each basal boron atom, then the equivalent molecular orbitals are

$a_1 \quad \psi_1 = \frac{1}{2}(\phi_1+\phi_2+\phi_3+\phi_4),$

$e \quad \begin{cases} \psi_2 = \dfrac{\phi_1-\phi_3}{\sqrt{2}}, \\[2mm] \psi_3 = \dfrac{\phi_2-\phi_4}{\sqrt{2}}, \end{cases}$

$b_1 \quad \psi_4 = \frac{1}{2}(\phi_1-\phi_2+\phi_3-\phi_4).$

The first three of these orbitals will interact with orbitals of like symmetry on the apical boron atom, generating three bonding orbitals and three antibonding orbitals. The b_1 orbital plays no part in bonding. Each of the sp^3 hybrid orbitals may be considered to contribute an electron and the apical boron two electrons. These six electrons can be housed in the three bonding orbitals.

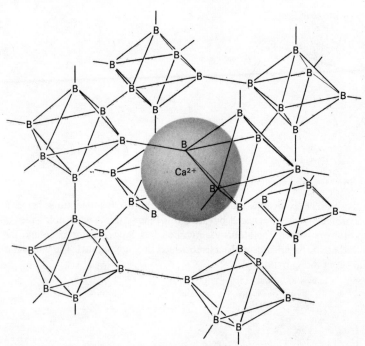

Figure 65   The structure of calcium boride CaB_6. The B_6^{2-} units are at the corners of cubes with the Ca^{2+} ions at the centres of the cubes

5.6.3   *Calcium boride,* CaB_6

Borides such as calcium boride are based on octahedral B_6^{2-} units joined by covalent bonds. If the octahedra are thought of as centred on the corners of a cube the bonds lie in the edges of the cube. This structure is repeated in three dimensions and the calcium cations are found at the centre of each cube (Figure 65). Whilst the interoctahedral boron–boron links can be regarded as simple localized bonds using sp hybrids from each boron atom the bonding within the B_6 unit is more complex. The representation of the six sp hybrids that point towards the centre of the octahedron (where there is no atom) and the twelve other p-orbitals are shown in Table 26.

Table 26

Orbitals	E	$8C_3$	$3C_2$	$6C_2'$	$6C_4$	i	$8S_6$	$3\sigma_h$	$6\sigma_v$	$6S_4$	
six sp hybrids	6	0	2	0	2	0	0	4	2	0	$a_{1g}+t_{1u}+e_g$
twelve p-orbitals	12	0	−4	0	0	0	0	0	0	0	$t_{1g}+t_{2g}+t_{1u}+t_{2u}$

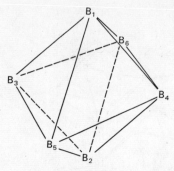

Figure 66   Octahedral structure of the B_6^{2-} anion

In the case of six ligands surrounding a central atom interactions between ligand orbitals could be ignored but here this interaction is what holds the six boron atoms together. The atoms are numbered as shown in Figure 66 and the p-orbitals are distinguished by the axis to which they are parallel. The molecular orbitals from the sp hybrids are

$$a_{1g} \quad \psi_1 = \frac{sp_1 + sp_2 + sp_3 + sp_4 + sp_5 + sp_6}{\sqrt{6}};$$

$$t_{1u} \quad \begin{cases} \psi_2 = \dfrac{sp_1 - sp_2}{\sqrt{2}}, \\[2mm] \psi_3 = \dfrac{sp_3 - sp_4}{\sqrt{2}}, \\[2mm] \psi_4 = \dfrac{sp_5 - sp_6}{\sqrt{2}}; \end{cases}$$

$$e_g \quad \begin{cases} \psi_5 = \dfrac{sp_1 + sp_2}{\sqrt{3}} - \dfrac{sp_3 + sp_4 + sp_5 + sp_6}{\sqrt{12}}, \\[2mm] \psi_6 = \tfrac{1}{2}(sp_3 + sp_4 - sp_5 - sp_6). \end{cases}$$

In order to find the energies of these molecular orbitals it is necessary to distinguish between end-on overlap of sp hybrids, β (e.g. sp_1 and sp_2) and sideways overlap, β' (e.g. sp_1 and sp_3). The energies of the molecular orbitals are

$$E(\psi_1) = \tfrac{1}{6}\{6(\alpha_{sp} + \beta + 4\beta')\}$$

$$= \alpha_{sp} + \beta + 4\beta',$$

$$E(\psi_2) = E(\psi_3) = E(\psi_4) = \tfrac{1}{2}\{2(\alpha_{sp} - \beta)\}$$

$$= \alpha_{sp} - \beta,$$

$$E(\psi_5) = E(\psi_6) = \tfrac{1}{4}\{4(\alpha_{sp} + \beta - 2\beta')\}$$

$$= \alpha_{sp} + \beta - 2\beta'.$$

The molecular orbitals from the p-orbitals are

$$t_{1g} \begin{cases} \psi_7 = \tfrac{1}{2}(p_{3y} - p_{4y} - p_{5x} + p_{6x}), \\[4pt] \psi_8 = \tfrac{1}{2}(p_{1x} - p_{2x} - p_{3z} + p_{4z}), \\[4pt] \psi_9 = \tfrac{1}{2}(p_{1y} - p_{2y} - p_{5z} + p_{6z}); \end{cases}$$

$$t_{2g} \begin{cases} \psi_{10} = \tfrac{1}{2}(p_{3y} - p_{4y} + p_{5x} - p_{6x}), \\[4pt] \psi_{11} = \tfrac{1}{2}(p_{1x} - p_{2x} + p_{3z} - p_{4z}), \\[4pt] \psi_{12} = \tfrac{1}{2}(p_{1y} - p_{2y} + p_{5z} - p_{6z}); \end{cases}$$

$$t_{1u} \begin{cases} \psi_{13} = \tfrac{1}{2}(p_{1x} + p_{2x} + p_{5x} + p_{6x}), \\[4pt] \psi_{14} = \tfrac{1}{2}(p_{1y} + p_{2y} + p_{3y} + p_{4y}), \\[4pt] \psi_{15} = \tfrac{1}{2}(p_{3z} + p_{4z} + p_{5z} + p_{6z}); \end{cases}$$

$$t_{2u} \begin{cases} \psi_{16} = \tfrac{1}{2}(p_{1x} + p_{2x} - p_{5x} - p_{6x}), \\[4pt] \psi_{17} = \tfrac{1}{2}(p_{1y} + p_{2y} - p_{3y} - p_{4y}), \\[4pt] \psi_{18} = \tfrac{1}{2}(p_{3z} + p_{4z} - p_{5z} - p_{6z}). \end{cases}$$

The most important interactions between the constituent p-orbitals of these molecular orbitals will be (a) those that lie in the edges of the octahedron, resonance integral β_1 (e.g. p_{1x} and p_{3z}) and (b) those derived from parallel orbitals on adjacent atoms, resonance integral β_2 (e.g. p_{3z} and p_{5z}). Other interactions will be ignored. The energies are then

$$E(t_{1g}) = 4(\alpha_p - 2\beta_2) = \alpha_p - 2\beta_2,$$

$$E(t_{2g}) = 4(\alpha_p + 2\beta_2) = \alpha_p + 2\beta_2,$$

$$E(t_{1u}) = 4(\alpha_p + 2\beta_1) = \alpha_p + 2\beta_1,$$

$$E(t_{2u}) = 4(\alpha_p - 2\beta_1) = \alpha_p - 2\beta_1.$$

Thus six bonding orbitals (t_{2g} and t_{1u}) and six antibonding orbitals (t_{1g} and t_{2u}) will be formed. If β and β' are comparable the sp hybrids will generate but one bonding orbital (a_{1g}). Molecular orbitals of representation t_{1u} are formed both from sp hybrids and also from the p-orbitals. Whilst interaction between these orbitals is probably small, overlap would lead to the formation of two new orbitals one more bonding and the other less bonding than either component orbital. The over-all situation of orbitals a_{1g}, t_{1u} and t_{2g} being bonding and e_g, t_{1u}, t_{2u}, t_{1g} being antibonding is unchanged. Thus fourteen electrons are required to hold the B_6 octahedron together. Each boron atom has committed one electron to forming a bond with a boron in a neighbouring octahedron so that the six boron atoms can only contribute twelve electrons; each B_6 unit needs two more electrons for stability. Hence a structure based on B_6^{2-}. The anion $B_6H_6^{2-}$ has recently been prepared, obviously it has the same basic electronic structure.

5.7 Resolution of a structural problem

The configuration of the dithionite anion $(S_2O_4^{2-})$ cis (C_{2v})–trans (C_{2h})

The dithionite anion consists of two SO_2^- units joined by a long weak sulphur–sulphur bond approximately at right angles to the planes of the SO_2^- fragments. Two configurations of some symmetry can be considered, the *trans* form in which oxygen pairs lie on opposite sides of the S–S bond as far from each other as possible (Figure 67a, C_{2h}) and the eclipsed *cis* form in which all the oxygen atoms are on the same side of the sulphur bond (Figure 67b, C_{2v}). Intuitively one might feel that the *trans* form would be preferred since here oxygen–oxygen repulsions and also sulphur–sulphur lone-pair repulsions would be minimized. Actually the *cis* form is observed, a result which can easily be understood if group theory is used to clarify the interactions between the various orbitals of importance. The account that follows is based on the argument presented by Dunitz ('The structure of sodium dithionite and the nature of the dithionite anion', *Acta Crystallographica*, vol. 9 (1956), pp. 579–86).

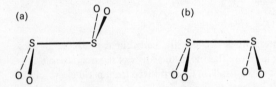

Figure 67   (a) *trans* (staggered) and (b) *cis* (eclipsed) forms of the dithionite ion $S_2O_4^{2-}$

Consider the first p-orbitals in an isolated SO_2^- unit at right angles to the plane of the three atoms. They will generate three molecular orbitals, one bonding, one nonbonding, relative to, and concentrated on, the oxygens, and one antibonding (cf. allyl π-orbitals, p. 33). The interaction of these orbitals upon the formation of dithionite, their symmetries in both *cis* and *trans* configurations and the electron occupancy of these orbitals is shown in Figure 68. Each sulphur atom also carries vacant $3d$ orbitals and the presence of electron-attracting oxygen ligands will enhance their bonding potential. The orbitals which play a critical role in bonding are those that overlap well with π-symmetry relative to, and in the region of, the S–S bond. One such pair (one orbital from each sulphur) lies in the plane that bisects both O–S–O angles (d_1s) and the other pair is in the plane at right angles, (d_2s). The molecular orbitals that result from the overlap of these d-orbitals are also shown in Figure 68.

The d_1 molecular orbitals belong to the same representations as some of the π molecular orbitals and so interaction between these orbitals is inevitable. In the *trans* ion the interaction between orbitals of b_u symmetry will be strongest and that between a_g orbitals will be less because the energy difference between

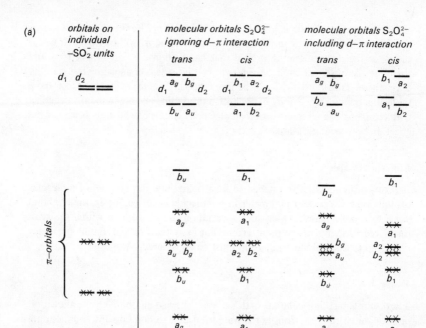

(a)

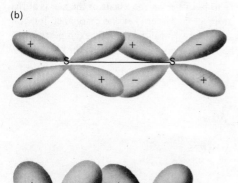

(b)

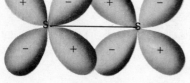

Figure 68 (a) Energy-level diagrams for molecular orbitals of the dithionite ion. (b) Overlap of sulphur d-orbitals

149 Resolution of a Structural Problem

the overlapping orbitals is greater for a_g than for b_u. For the same reason a_1 interaction will be strong in the *cis* configuration and b_1 will be weak. Now it is the a_g or the a_1 antibonding orbital that has to house a pair of electrons and therefore the observed configuration will be that which confers greatest stability upon them. Thus *cis*-$S_2O_4^{2-}$ will be more stable than *trans*-$S_2O_4^{2-}$. Since the p molecular orbitals a_u, b_g or a_2, b_2 are both nonbonding and are both fully occupied with electrons, the interaction of d_2 molecular orbitals has an equally stabilizing effect on both configurations.

5.8   Molecular shapes

It is often of considerable interest to investigate how the energies of molecular orbitals vary if a certain parameter in a molecule is varied, for example a bond angle or a bond length. Often such a variation can be allowed within the symmetry requirements of the point group, but in systems of high initial symmetry reduction to a less symmetric group must be considered.

5.8.1   *Water, H_2O*

The discussion at the beginning of this chapter showed qualitatively the allocation of the eight valence electrons in the water molecule to bonding or to nonbonding orbitals. If we make further assumptions about the way in which the resonance integrals β_1, β_2 and β_3 vary with the H–O–H angle (2θ) then it will be possible to see how the energies of the molecular orbitals vary with bond angle.

Suppose hydrogen atom H_a were to lie either on the x-axis or the z-axis at the same distance from the oxygen atom as in water. Let the resonance integral $\int (H_a) \mathscr{H} p_x$ (or p_z) $d\tau = \beta$. As H_a moves away from either axis, for example through θ from the z-axis, let us assume the resonance integral varies as the cosine of θ.

$$\text{Thus} \quad \beta_2 = \int \psi_1 \mathscr{H} p_z. d\tau = \frac{1}{\sqrt{2}} \int (H_a + H_b) \mathscr{H} p_z \, d\tau$$

$$= \frac{2}{\sqrt{2}} \beta \cos \theta$$

$$= \beta \sqrt{2} \cos \theta.$$

$$\text{Similarly} \quad \beta_3 = \frac{2}{\sqrt{2}} \beta \sin \theta$$

$$= \beta \sqrt{2} \sin \theta.$$

If we further assume $\alpha_H = \alpha_p = \alpha$ and write $(\alpha - E)/\beta = x$ then the b_1 determinant becomes

$$\begin{vmatrix} x & \sqrt{2} \sin \theta \\ \sqrt{2} \sin \theta & x \end{vmatrix} = 0,$$

i.e. $\quad\quad\quad\quad x = \pm \sqrt{2} \sin \theta.$

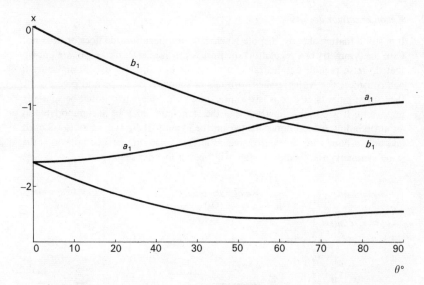

Figure 69   To show variation of lowest bonding orbital energies with θ for the water molecule. The two other electron pairs are housed in nonbonding orbitals (a_1 and b_2). Lowest curve represents total energy as function of θ

Also the a_1 determinant may be written

$$\begin{vmatrix} x+\varepsilon & 0 & 1 \\ 0 & x & \sqrt{2}\cos\theta \\ 1 & \sqrt{2}\cos\theta & x \end{vmatrix} = 0.$$

Here it has been assumed that $\beta_1 = \beta_3$ and α_s has been expressed as equal to $\alpha_p + \varepsilon\beta_1$. If $\varepsilon = 0$ the roots are $x = 0$ and $x = \pm\sqrt{(1+2\cos^2\theta)}$. The a_1 and b_1 molecular orbital energies as a function of θ, with all these assumptions, are plotted in Figure 69. It can be seen that the sum of the bonding-orbital energies has its largest negative value (i.e. most bonding) when $\theta = 60°$ (i.e. bond angle 120°), but it can also be seen that the total energy varies little between $\theta = 45°$ and $\theta = 90°$ suggesting that quite small factors may cause substantial changes in bond angle. For water probably the most erroneous assumption made above was that $\varepsilon = 0$. Actually the difference between α_s and α_p for oxygen is about 15 eV so that ε might well be ~ 2. In this particular case the most stable value for θ is $\sim 50°$. In general as ε increases the s-orbital plays a less important role in bonding. When this happens the a_1 determinant may be simplified by omitting the s-orbital altogether and the most stable bonding condition is achieved at $\theta = 45°$, bond angle 90°.

It is still a matter of some dispute whether xenon hexafluoride does or does not have the symmetry of a regular octahedron. Many isoelectronic anions are known that do have regular octahedral symmetry, for example $TeCl_6^{2-}$, $SeBr_6^{2-}$ etc. If one considers the 'valence-shell' orbitals of xenon to be $5s$, $5p$ and possibly $5d$, then seven pairs of electrons (six bonding and one lone pair) should be spatially important. It is possible however that the 'lone pair' may be accommodated in an antibonding orbital concentrated on the ligands. If O_h symmetry is assumed and only σ-bonds are considered then as in the transition metal complexes of the same symmetry the ligand orbitals will belong to representations $a_{1g} + t_{2u} + e_g$.

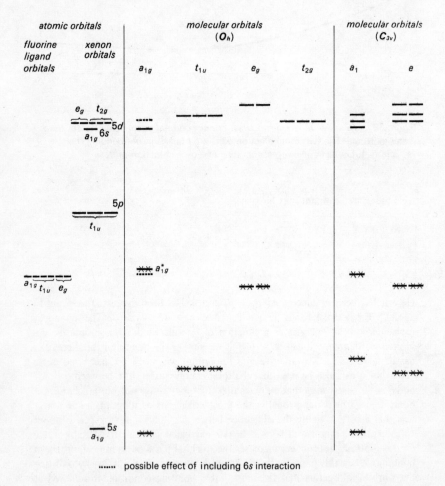

Figure 70   Qualitative energy-level diagram for xenon hexafluoride

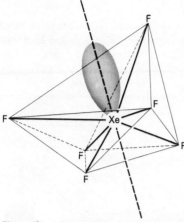

Figure 71   Distorted octahedral structure for xenon hexafluoride

On xenon, orbitals of corresponding symmetry are $5s(6s) - a_{1g}$; $5p$ orbitals $- t_{2u}$; $d_{x^2-y^2}$ and $d_{z^2} - e_g$. A very qualitative energy-level diagram is shown in Figure 70. Since the $5s$ xenon orbital is rather tightly bound it will not contribute much to the bonding and the a_{1g}^* antibonding orbital, which will be almost exclusively concentrated on the ligands, will not be very strongly antibonding. It is therefore quite reasonable that the seventh pair of electrons could be accommodated here and regular octahedral symmetry retained.

Suppose some distortion were to occur, so that the lone pair were to concentrate in the middle of one of the faces of the octahedron. Taking this as the site of the new z-axis, the three fluorine atoms nearest the lone pair will move away and in turn cause the other three fluorine atoms to come nearer to the z-axis, at the opposite end from the lone pair (Figure 71). The symmetry of the distorted structure is only C_{3v}. In order to correlate the point groups, find those operations in O_h that correspond to those of C_{3v}: E, C_3 and σ_v. (Examine both σ_v and σ_h in O_h carefully – it will then be seen that σ_v planes in O_h are in fact the same as σ_v in C_{3v}.) Now read off the characters in O_h for these three operations and so find the representations in C_{3v}. Triply degenerate representations in O_h will be reducible in C_{3v}. Thus

O_h	C_{3v}	O_h	C_{3v}
A_{1g} and $A_{2u} \rightarrow A_1$,		T_{1g} and $T_{2u} \rightarrow E + A_2$,	
A_{1u} and $A_{2g} \rightarrow A_2$,		T_{1u} and $T_{2g} \rightarrow E + A_1$.	
E_g and $E_u \rightarrow E$,			

One effect of this reduction of symmetry would be to permit the d_{xy} and d_{yz} orbitals to participate, in principle at least in the bonding. It is conceivable that this interaction could stabilize the ex-a_{1g}^* orbital and so make the distorted structure more stable. This argument does not resolve the problem but does provide a framework for a discussion of the energetics of distortion.

5.8.3 Flattening methane, $CH_4: T_d \rightarrow D_{2d} \rightarrow D_{4h}$

The intermediate form of methane in which the tetrahedron has been somewhat flattened has symmetry D_{2d}. The representations of the carbon and hydrogen valency orbitals are shown in Table 27.

Table 27

Orbitals	E	C_2	$2S_4$	$2C_2'$	$2\sigma_d$	
four hydrogens	4	0	0	0	2	a_1+e+b_2
carbon $2s$	1	1	1	1	1	a_1
carbon $2p_z$	1	1	-1	-1	1	b_2
carbon $2p_x$ and $2p_y$	2	-2	0	0	0	e

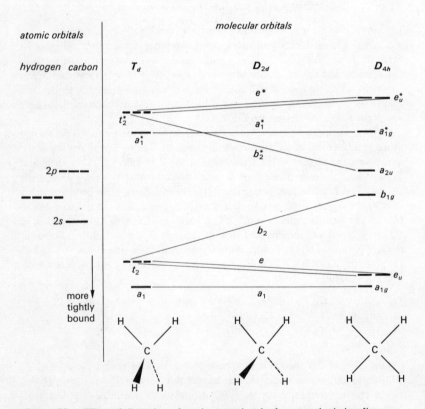

Figure 72   Effect of distortion of methane molecule, from tetrahedral to flat, on the energy levels of the molecular orbitals

The symmetry representations may be carried through from the tetrahedral group to D_{2d} but t_2, the representation for p_x, p_y and p_z, becomes $e+b_2$. Since the hydrogen atoms are approaching the xy plane and falling away from the z-axis it is reasonable to suppose that the resonance integrals for the interaction of e carbon and hydrogen orbitals will increase but that the integral for b_2-type overlap will decrease (these integrals are, of course, identical in T_d). The effect of this distortion upon the energy levels of the molecule is shown qualitatively in Figure 72.

In the extreme case – flat CH_4 – the representations are as in Table 28.

Table 28

Orbitals	E	C_2	$2C_4$	$2C_2'$	$2C_2''$	i	σ_h	$2S_4$	$2\sigma_v'$	$2\sigma_v''$	
four hydrogens	4	0	0	2	0	0	4	0	2	0	$a_{1g}+e_u+b_{1g}$
carbon $2s$	1	1	1	1	1	1	1	1	1	1	a_{1g}
carbon $2p_z$	1	1	1	-1	-1	-1	-1	-1	1	1	a_{2u}
carbon $2p_x$ and $2p_y$	2	-2	0	0	0	-2	$+2$	0	0	0	e_u

Such a structure would be held together by only three bonding orbitals derived from interactions between carbon and hydrogen orbitals a_{1g} and e_u. The carbon $2p_z$ (a_{2u}) orbital and the b_{1g} hydrogen molecular orbital would both be non-bonding.

Thus the regular tetrahedral shape gives the greatest scope for accommodating eight electrons in bonding orbitals. If methane were ionized, CH_4^+, the remaining seven electrons may be housed, two in the most tightly bound orbital a_1 and five in the t_2 orbitals.

Thus the electron population in each of the component orbitals of the t_2 set cannot be identical. But the t_2 orbitals in T_d are equivalent. This anomalous situation may be resolved: the molecule distorts to a lower symmetry (Jahn–Teller effect) in which the 't_2' orbitals are no longer necessarily degenerate. This is the case for flattened CH_4^+. Here the electrons can be accommodated (in orbitals of decreasing bonding power), a_1^2, e^4, b_1^1. Exactly how flat CH_4^+ would go will depend on the relative sizes of the e and b_2 resonance integrals for various angles and upon electron repulsion-correlation effects, etc.

5.9 Symmetry conservation in chemical reactions

Chemical reactions also provide examples of systems in which the symmetry of the original molecule is often considerably reduced. In order to follow the course of a reaction in detail it is of great interest to study the way in which each molecular orbital behaves throughout the reaction. If the various stages of the reaction are characterized by symmetry then the problem of correlating particular orbitals

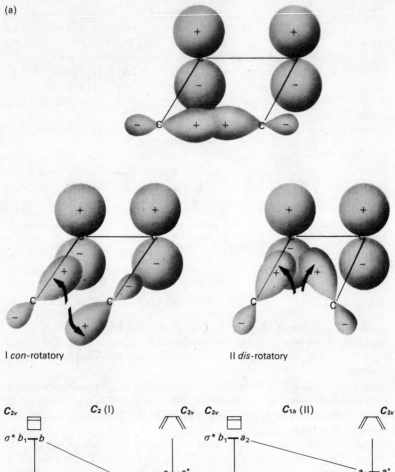

Figure 73 Rearrangement of cyclobutene to *cis*-buta-1,3-diene: (a) orbital diagram of alternative reaction processes; (b) energy-level diagrams for the reaction

during the reaction is considerably simplified. A most interesting example of this type of study concerns reactions involving olefins and the thermal or photochemical reactions in which σ-bonds and π-bonds are interconverted. The symmetry point group used will be that relevant to the bonds participating in the reaction rather than that of the whole molecule. It is reasonable to assume that groups not directly taking part in the reaction will not affect its course materially.

Let us consider the rearrangement of the cyclobutene system to the isomeric buta-1,3-diene. If we assume suitable hybridization then the bond opposite the double bond in cyclobutene will arise from the overlap of two sp^3 hybrids. For simplicity the p-character will be emphasized – Figure 73 – in which two possible routes to the product molecule are considered, I con-rotatory and II dis-rotatory. Either mechanism assumes that the breaking of the σ-bond and its rearrangement to the conjugated π-system is concerted. Clearly the C_{2v} symmetry of cyclobutene is destroyed; I belongs to C_2 and II to C_{1h}. The product molecule cis-buta-1,3-diene has again C_{2v} symmetry. By considering only symmetry correlations and remembering that orbitals of the same representation will interact and therefore not cross each other on an energy level diagram it is possible to construct continuous relationships between initial and final molecular orbitals. If electronic excitation of the π-bond were to initiate the reaction an orbital of a_2 symmetry would be populated. It is immediately obvious that route I is energetically most unfavourable compared with II. In the former case the final configuration in the butadiene could be $b_2^2 a_2^1 b_2^{*0} a_2^{*1}$ whereas by route II the more stable $b_2^2 a_2^1 b_2^{*1} a_2^{*0}$ would result. Conversely if no electronic excitation takes place – that is if the reaction is started by heat – then route I is favoured, leading to $b_2^2 a_2^2 b_2^{*0} a_2^{*0}$. Route II leads to $b_2^2 a_2^0 b_2^{*2} a_2^{*0}$.

Thus the reaction route depends upon the way in which the reaction is carried

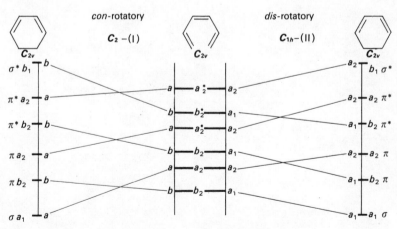

Figure 74   Energy-level diagrams for alternative reaction paths for the rearrangement of cyclohexa-1,3-diene to hexa-1,3,5-triene

out. If suitable marker groups are attached to the cyclobutene system it is possible to distinguish the products and so test the theory.

The very same type of arguments summarized in Figure 74 show that photochemical and thermal reactions follow opposite routes if the starting material is cyclohexa-1,3-diene rearranging to hexa-1,3,5-triene. The alteration of reaction routes with number of double bonds is quite general.

In a similar way, Diels–Alder reactions between systems with even and odd numbers of double bonds under thermal conditions can be understood. The presence of ultraviolet light permits even–even and odd–odd π-bond systems to couple.

Problems 5

5.1 *Non-equivalent hybridization.* Suppose the $2s$ and $2p$ orbitals of carbon have been used to make *two* sp^2 hybrids. Use the remaining portions of the atomic orbitals to make two other (equivalent) hybrid orbitals.

5.2 For f-orbitals, $l = 3$. Use the equation on p. 170 to determine the irreducible representations of a set of seven f-orbitals under T_d symmetry. Could these orbitals form π-bonds with oxygen lone pairs in IO_4^- or TeO_4^{2-}?

Chapter 6
Visible and Ultraviolet Spectroscopy

When electromagnetic radiation is absorbed by atoms or molecules it promotes them to an excited state. Microwave and infrared radiation correspond to rather low energy quanta and so initiate rotational and vibrational excitation. Visible and ultraviolet light on the other hand (λ, 100–800 nm) have much higher frequencies (in wave numbers 10 000–1250 mm^{-1}) and can cause excitations of the order of 12·5–1·5 eV per molecule. Such energies are characteristic of electronic excitation: the promotion of an electron from one orbital to another. The energy required to effect this transition is of course the energy absorbed. Thus we expect molecules to absorb light of particular wavelengths corresponding to the energy differences between occupied and unoccupied orbitals. The amount of energy absorbed varies greatly with bond type. In strong σ-bonds between first-row elements the electrons are very much more tightly bound than in the free atoms. Electronic excitation to the antibonding orbital σ^* will therefore require a great deal of energy, typically 10–12 eV for bonds such as C–C, C–H, N–H, O–H, etc. In many molecules that contain such bonds some electrons are still in atomic orbitals – 'lone pairs'. The energy required to excite electrons from these orbitals to σ-orbitals, symbolized $n \rightarrow \sigma^*$, is less than for $\sigma \rightarrow \sigma^*$ transitions; it is often of the order 6–8 eV. In organic molecules π bonds are weaker bonds than σ-bonds so that π-orbitals and π^*-orbitals are nearer in energy and $\pi \rightarrow \pi^*$ transitions can easily be observed at energies of less than 5–6 eV. Often $n \rightarrow \pi^*$ transitions take place at even lower energies.

In all these examples the energy difference between bonding and antibonding orbitals has been such that in the ground state the electrons have entered the bonding orbitals in pairs. If the interaction between the original atomic orbitals had been less this energy difference would have been less also. When this is so, the energy lost by electron repulsion upon pairing off two electrons in one bonding orbital is more than the energy lost by putting one electron of the pair into the antibonding orbital. The latter situation is more stable and is the ground state. This sort of weak interaction is found in many transition metal complexes.

6.1 Selection rules

Let us now look in some detail at the symmetry properties of the orbitals between which electronic transitions take place. Visible and ultraviolet light are electromagnetic radiations and it is the interaction of the electric vector with molecules (or atoms) that causes the most intense absorption. Indeed the intensity of absorption can be calculated by evaluation of the integral

$$p = \int \Psi_g \mathscr{P} \Psi_{ex},$$

where p is the probability that the transition from Ψ_g to Ψ_{ex} will take place, Ψ_g is the molecular wave function for the ground state, Ψ_{ex} is the molecular wave function for the excited state and $\mathscr{P}$ is the operator initiating the transition from Ψ_g to Ψ_{ex}. This integral is of course very similar in form to the one we met at the beginning of the last chapter. This equation too must balance in symmetry.

Thus $\quad \Gamma(p) = \Gamma(\Psi_g) \Gamma(\mathscr{P}) \Gamma(\Psi_{ex}).$ **6.1**

Now p is a scalar number, a probability, and so $\Gamma(p)$ must be wholly symmetric. If we are considering only transitions brought about by the electric vector of the radiation then $\Gamma(\mathscr{P})$ will be the same as the representations of the electric vectors; that is the same as $\mathbf{x}$, $\mathbf{y}$ and $\mathbf{z}$ (which are of course the same as p_x, p_y and p_z). When the right-hand side of **6.1** does not yield the wholly symmetric representation then p is necessarily zero and the transition is said to be *forbidden*: the anticipated transition is not observed. When the evaluation of the direct product in **6.1** does give the wholly symmetric representation the transition is said to be *allowed* although this does not tell you whether p will be 0·001 or 0·3 or even 1·0; that is it gives no indication of how intense the absorption will be, merely that there is no symmetry prohibition on its taking place. Even so, molecular symmetry is particularly useful in classifying and interpreting ultraviolet and visible absorption spectra.

The representations of the electronic wave functions Ψ_g and Ψ_{ex} may be found by considering the representations of the individual molecular orbitals. The representation of an electron in an orbital is simply the representation of that orbital. The representation for an over-all molecular configuration is the direct product of the representations for each electron. Thus the configuration $\psi_1^2 \psi_2^2 \psi_3^1 \psi_4^1$ has the representation

$$\Gamma(\psi_1) \Gamma(\psi_1) \Gamma(\psi_2) \Gamma(\psi_2) \Gamma(\psi_3) \Gamma(\psi_4).$$

Notice that doubly occupied (or empty) non-degenerate orbitals yield the wholly symmetric representations so that the above product can be more simply evaluated as

$$\Gamma(\psi_3) \Gamma(\psi_4).$$

If all the electrons in the ground state of a molecule are paired off so that no orbital is singly occupied then the representation Ψ_g is wholly symmetric.

In practice it will be found particularly easy to determine whether a transition is permitted or forbidden by using the symmetry properties of the molecule.

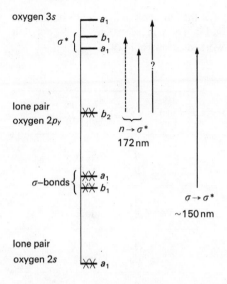

oxygen 3s — a_1

σ^* { — b_1 / — a_1

?

lone pair oxygen $2p_y$ — b_2

$n \rightarrow \sigma^*$
172 nm

σ-bonds { — a_1 / — b_1

$\sigma \rightarrow \sigma^*$
~150 nm

lone pair oxygen 2s — a_1

Figure 75   Energy-level diagram for molecular orbitals of the water molecule showing some possible electron transitions (cf. Figure 50)

6.1.1   *Examples*

Water. The array of valence molecular orbitals is shown diagrammatically in Figure 75. The longest wavelength transition will be from the nonbonding b_2 orbital to an antibonding b_1 or a_1 orbital or even to a 3s orbital on the oxygen atom (a Rydberg transition). $\Gamma(\Psi_{ex})$ for two possible transitions would be

$$b_2 b_1 = a_2 \quad \text{or} \quad b_2 a_1 = b_2.$$

The former would be forbidden but the latter permitted since $\Gamma(\mathscr{P}) = a_1, b_1$ or b_2. A strong transition is observed in water with a maximum at about 172 nm, in the vacuum ultraviolet. Even so, this is a longer wavelength than for $\sigma \rightarrow \sigma^*$ transitions (e.g. for C–H bonds), in keeping with the interpretation that the transition is from a nonbonding orbital to an antibonding orbital: $n \rightarrow \sigma^*$. A second absorption is observed in water in the vacuum ultraviolet at a wavelength shorter than 150 nm, this presumably is a $\sigma \rightarrow \sigma^*$ transition.

Cis- and trans-*buta*-1,3-*diene* (π-*orbitals*), C_{2v} *and* C_{2h}. The nature of the four π molecular orbitals in both *cis*- and *trans*-butadiene can easily be found (see p. 131). The representations of the electronic ground states are $\Gamma(cis) = b_2^2 a_2^2 = A_1$ and $\Gamma(trans) = a_u^2 b_g^2 = A_g$.

Since to a first approximation the energies of the molecular orbitals are the same in both conformations the electronic excitations in both should have similar energies – it will be of interest to determine whether the shape of the molecule has any effect upon their intensities. Let the four molecular orbitals be, in order of decreasing energy, ψ_1, ψ_2, ψ_3 and ψ_4. The representations for Ψ_{ex} for four simple one-electron excitations are worked out in Table 29.

Table 29

Excited states	Representations of occupied m.o.s	$\Gamma(\Psi_{ex})$	Electric vector to give transition
$\psi_1^2 \psi_2^1 \psi_3^1$	cis $b_2^2 a_2 b_2$	B_1	b_1
	trans $a_u^2 b_g a_u$	B_u	b_u
$\psi_1^2 \psi_2^1 \psi_4^1$	cis $b_2^2 a_2 a_2$	A_1	a_1
	trans $a_u^2 b_g b_g$	A_g	none
$\psi_1^1 \psi_2^2 \psi_3^1$	cis $b_2 a_2^2 b_2$	A_1	a_1
	trans $a_u b_g^2 a_u$	A_g	none
$\psi_1^1 \psi_2^2 \psi_4^1$	cis $b_2 a_2^2 a_2$	B_1	b_1
	trans $a_u b_g^2 b_g$	B_u	b_u

Since $\Gamma(\Psi_g) = A_1$ or A_g the direct product of $\Gamma(\mathscr{P})$ and $\Gamma(\Psi_{ex})$ must be wholly symmetric for the transition to be permitted. $\Gamma(\mathscr{P})$ for *cis*-butadiene is $a_1(z)$, $b_1(x)$, $b_2(y)$ and for *trans*-butadiene it is $a_u(z)$, $b_u(x$ and $y)$; the component of $\mathscr{P}$ that would make the transition from the ground state permitted is indicated in the last column of Table 29.

Thus whilst the long-wave transition to $\psi_1^2 \psi_2 \psi_3$ is permitted for both *cis* and *trans* molecules the higher energy transitions to $\psi_1^2 \psi_2 \psi_4$ and $\psi_1 \psi_2^2 \psi_4$ should be observed only for the *cis* isomer. Unfortunately this distinction is difficult to observe experimentally since both absorptions are at rather short wavelengths. It has been suggested that the first is at about 210 nm and the second at about 175 nm, just in the vacuum ultraviolet. A further difficulty is that whilst it is easy to obtain molecules in which the double bond system must be *cis* (e.g. cyclopentadiene) wholly *trans* molecules are rare.

Benzene, C_6H_6. The π molecular orbitals of benzene are a_{1u}, e_{1g} (bonding), e_{2u}, a_{2g} (antibonding). The long-wave electronic transition will be associated with excitation from the e_{1g} orbitals to e_{2u} orbitals. The representation of e_{1g}^3 will be the same as for e_{1g}^1, since the symmetry properties of an electron vacancy ('electron hole') are the same as for an electron itself. Also $\Gamma(e_{2u}^1) = e_{2u}$. The representation for Ψ_{ex} is therefore $e_{1g}e_{2u} = b_{1u} + b_{2u} + e_{1u}$. These three states are to a first approximation energetically degenerate but the degeneracy will be removed by electronic interaction (which will be important in electronic excitation).

The ultraviolet spectrum of benzene shows therefore three quite different absorption frequencies. $\Gamma(\mathscr{P})$ is $a_{2u} + e_{1u}$ so that the transition to b_{1u} and b_{2u} states are forbidden but to e_{1u} is permitted – $e_{1u}e_{1u} = a_{1g} + a_{2g} + e_{2g}$, the product contains the wholly symmetric representation. Extensive calculations by Goeppert-Mayer and Sklar ('Calculations of the lower excited levels of benzene', *Journal of Chemical Physics*, vol. 6 (1938) no. 10, pp. 645–52) and later authors have suggested that the collection of weak bands between 240 and 265 nm is to be associated with the forbidden transition to the B_{2u} state. These bands are observed because certain vibrational frequencies are stimulated simultaneously with the electronic excitation. In order to determine which vibrations can be called into play it is necessary to know their symmetry properties (Chapter 7). The basic equation **6.1** still holds. The wave functions should now not only include electronic symmetry factors but also those for the vibrations, thus,

$\Gamma(p)$ = wholly symmetric

$\quad$ = $\Gamma(\Psi_g \text{ electronic})\,\Gamma(\Psi_g \text{ vibration})\,\Gamma(\mathscr{P} \text{ electric vectors})\,\Gamma(\Psi_{ex} \text{ electronic})$
$\quad\quad \Gamma(\Psi_{ex} \text{ vibration})$

Obviously if no vibrational excitation takes place then

$\Gamma(\Psi_g \text{ vibration}) = \Gamma(\Psi_{ex} \text{ vibration})$,

and the equation reduces to equation **6.1**. If we assume the molecule to be in its ground vibrational as well as ground electronic state the problem reduces itself to finding which vibrations have representations that can be used to multiply the direct product of $\Gamma(\mathscr{P}) \times \Gamma(\Psi_{ex} \text{ electronic})$ to give a wholly symmetric representation. These particular vibrations would then be said to couple with the electronic excitation to make it permitted – *vibronic coupling*. Of course the actual frequency of absorption will be the sum of the frequency of the pure electronic transition and the frequency of the vibration(s) excited.

In this example $\Gamma(\Psi_{ex} \text{ electronic})$ is B_{2u} and $\Gamma(\mathscr{P})$ is $a_{2u} + e_{1u}$ so that $\Gamma(\Psi_{ex} \text{ vibration})$ must be either b_{1g} or e_{2g} to make the transition permitted by vibronic coupling. ($b_{2u}a_{2u}b_{1g} = a_{1g}$ and $b_{2u}e_{1u}e_{2g} = a_{1g} + a_{2g} + e_{2g}$). Unfortunately benzene possesses no B_{1g} vibrations so that only the E_{2g} vibration and its odd overtones will be able to make this electronic transition permitted. An interesting complication now arises, for benzene possesses A_{1g} vibrations and they could combine with the B_{2u} (electronic), E_{2g} (infrared) state to generate other excited states to which transitions from the ground state would be possible. Of course transitions to the B_{2u} electronic state would be impossible if aided only by the A_{1g} vibration. This is in fact what is observed. Since there are four E_{2g} and two A_{1g} vibrations it can be seen that the array of weak vibrationally permitted electronic transitions will be quite numerous. Furthermore a wide variety of other pairs of vibrations (many infrared inactive) can facilitate the transition to the B_{2u} state (see Ingold *et al.*, 'Excited states of benzene. Parts I–XII', *Journal of the Chemical Society*, 1948, pp. 406–517).

Groups of atoms such as $>C-O$, $=NO_2$ retain much of their spectral character even upon incorporation into larger molecules. Wavelengths of absorption are usually altered but the appearance of the absorption bands is surprisingly constant. The carbonyl group can often be associated with a broad weak absorption at about 280–300 nm and the nitro group almost invariably introduces colour into a compound, especially if π-conjugation is also present.

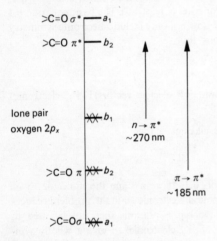

Figure 76   Energy-level diagram for orbitals of the carbonyl group showing electron transitions

6.2.1   *Carbonyl*

Assume that the local symmetry is C_{2v}, that the carbon is sp^2 hybridized and that the oxygen 2s orbital is so tightly bound that it may be ignored in this discussion. Let the orbitals be orientated as in Figure 76. The carbon sp^2 hybrid pointing at the oxygen and the oxygen p_z orbital have a_1 symmetry and will form σ and σ^* orbitals. Similarly the carbon p_y and the oxygen p_y will form π and π^* orbitals of b_2 symmetry. Oxygen $p_x(b_1)$ will be nonbonding. The lowest energy electronic transition will be from the b_1 to the $\pi^*(b_2)$ orbitals, $\Gamma(\Psi_{ex}) = b_1 b_2 = a_2$. Such a transition is therefore forbidden. This very simple treatment has ignored the disturbing effect that other groups attached to the carbonyl group may have upon it, and also the effects of molecular vibration. Either or both could easily lead to the $n \rightarrow \pi^*$ transition being weakly allowed. Conjugation results in the proliferation of both bonding and antibonding orbitals with a consequent lowering of the $n \rightarrow \pi^*$ energy gap so that the absorption moves to a longer wavelength –

bathochromic shift. The $\pi \to \pi^*$ transition in the carbonyl group yields

$$\Gamma(\Psi_{ex}) = b_1 b_1 = a_1$$

and is permitted. Formaldehyde vapour, for example, has a strong absorption at about 185 nm.

6.2.2 Nitro group

Again let us assume a local symmetry of C_{2v} but consider the π-orbitals. The nitrogen $2p$ belongs to representation b_2 but the two oxygen $2p$ orbitals to a_2 and b_2. The b_2 orbitals will form bonding (π) and antibonding (π^*) orbitals but the a_2 orbital will be nonbonding (π^0) and localized on the oxygen atoms. The long wavelength absorption between π-orbitals will arise from the $\pi^0 \to \pi^*$ transition, for which $\Gamma(\Psi_{ex}) = a_2 b_2 = b_1$. Each oxygen atom also carries a lone pair of electrons in nonbonding orbitals in the plane of the atoms: these orbitals have representations a_1 and b_1. Thus there will be two possible excited states arising from an $n \to \pi^*$ transition, either $\Gamma(\Psi_{ex}) = a_1 b_2 = b_2$ or $\Gamma(\Psi_{ex}) = b_1 b_2 = a_2$. The former is permitted: the latter forbidden. The allocation of transitions is not yet definite but it is tempting to associate the weak long-wave absorption of nitro-compounds with the forbidden $n \to \pi^*$ transition and the strong absorption at shorter wavelengths either with the permitted $n \to \pi^*$ or with the $\pi^0 \to \pi^*$ transition. As with carbonyl compounds π-conjugation with larger systems leads to lower-lying π^*-orbitals and a consequent bathochromic shift of the 'nitro' bands.

6.3 Crystal field theory

Interaction between the constituent atoms of a molecule is not always as strong as in the examples considered above. The bonding in transition metal complexes is often much weaker. In many cases an understanding of the ultraviolet spectrum may be had by assuming that the central atom or ion is not greatly altered by the bonding with the ligands. Such complexes may be regarded as a central atom or ion subject to a perturbing field whose symmetry is the arrangement of ligand orbitals pointing towards the atom; for an octahedral complex the field would have O_h symmetry. Since this situation is very similar to that for ions in a crystal, crystal field theory, which describes how the energies and degeneracies of gaseous atoms and ions are altered in a crystal environment, may be extended to encompass transition metal complexes. It will be convenient to discuss first the electronic structures of atoms and then describe the effect of the crystal field upon them.

6.3.1 Electronic states of atoms

Electrons may be arranged, even in one atom, in a variety of different ways giving rise to various different electronic configurations. These configurations may be written as the number of electrons with specific n and l quantum numbers, for example an unexcited carbon atom, $1s^2 2s^2 2p^2$. This description is not unique because the p-orbitals ($l = 1$) differ in their magnetic quantum numbers m_l and also the electrons differ in their spins, $m_s = \pm\frac{1}{2}$. The m_l quantum numbers may be added together for each of the electrons to give an analagous atomic quantum number M_L. For the two p-electrons in carbon M_L could have the values $+2$ ($m_{l_1} = +1, m_{l_2} = +1$), $+1$ ($m_{l_1} = +1, m_{l_2} = 0$, or $m_{l_1} = 0, m_{l_2} = +1$), 0, -1 or -2. These values for M_L describe possible orientations of the atomic angular-momentum quantum number L in a magnetic field. As with the orbital quantum number l, there are $2L+1$ possible values of M_L associated with a given L: $+L$, $+(L-1)$, ..., $-L$. The individual spin quantum numbers m_s are coupled in a similar manner to yield, for any particular arrangements of electron spins, a quantum number M_S. These values M_S are components of the atomic electron-spin quantum number S; associated with a given S there are $2S+1$ possible values of M_S: $+S$, ..., $-S$.

Atomic terms are characterized by specific values L and S, and have quite different energies (differences typically of the order of an electron volt or so) even for the same configuration, for different values of L and S.

Russell-Saunders or L–S coupling. The atomic quantum numbers L and S themselves may couple to give an over-all atomic angular-momentum quantum number J; the permitted values of J range from $L+S$, $L+S-1$ to $L-S+1$, $L-S$ if $L > S$ or from $S+L$, $S+L-1$ to $S-L$ if $L < S$. In the former case there will be $2L+1$ values of J, and in the latter $2S+1$ values of J, but in either case the spin is given by $2S+1$. This 'spin–orbit' coupling introduces, for a given L and a given S, only very small energy change so that the energy levels characterized by different Js typically differ by only about 10^{-2}–10^{-4} eV for the lighter atoms of the periodic table.

Since J is a quantum number that describes angular momentum there will be $2J+1$ possible orientations in a magnetic field; that is in the absence of a magnetic field the energy level characterized by a specific J is $2J+1$ degenerate. The situation is analogous to that for simple orbital angular momentum – d-orbitals have $l = 2$ so that there are $2l+1 = 5$ d-orbitals all of which have the same energy in the absence of a magnetic field.

The orange colour of a sodium flame is mainly due to an electronic transition from a $3s$ to a $3p$ orbital. Concentrating just upon this electron, L and S for $3s^1$ are 0 and $\frac{1}{2}$ respectively so that $J = \frac{1}{2}$ but for $3p^1$, $L = +1$, $S = +\frac{1}{2}$ so that J may be either $+\frac{1}{2}$ or $+\frac{3}{2}$. The selection rules for permitted electronic transitions are $\Delta J = 0$ or $\Delta J = \pm 1$ so that both possible $3p^1$ levels will be populated. This gives rise to the sodium 'doublet', two absorption lines at 588·9 and 589·5 nm. In working out the possible terms for atoms one may ignore all closed shells since $\sum m_l = 0 = L$ and $\sum m_s = 0 = S$; the contribution from such shells to the over-all atomic L and S will be zero.

The symbolism for describing atomic states is $^{2S+1}L_J$. Thus the ground state of the sodium atom is $^2S_{\frac{1}{2}}$ and the two possible $3p^1$ energy levels are $^2P_{\frac{1}{2}}$ and $^2P_{\frac{3}{2}}$. Capital letters are used for various values of L, thus 0, S; 1, P; 2, D; 3, F; 4, G, etc.

j–j coupling. The coupling between orbital and electron functions described above holds quite well for the valence and outer shells of light atoms (i.e. up to about chlorine). It is found, however, that the degree of coupling between L and S increases steadily with atomic number causing greater and greater energy differences between states with the same L and S but different J. Another way of expressing this is to say that L and S are no longer good quantum numbers. A complete breakdown of L–S coupling is envisaged in the *j–j* coupling scheme. Here *j* is evaluated for each electron; *j* can have values $l+s, l+s-1, ..., l-s$ and the *j*s couple to give J. Under this scheme values of L and S have no meaning. This extreme case is not reached but the spectra of heavier elements can only be satisfactorily interpreted by assuming a coupling somewhat transitional between L–S and *j–j*. It should be noted that L–S nomenclature is often used even when this type of coupling is no longer strictly applicable.

For the remainder of this discussion L–S coupling will be assumed.

Microstates. Returning to the carbon atom it is now possible to evaluate the various possible energy levels of the atom for different arrangements of the $2p^2$ electrons. The Pauli exclusion principle that no two electrons may have the same four quantum numbers places certain limitations upon the possible arrangements or microstates. The term *microstates* is used to describe a group of electrons each one of which has a specific m_l and m_s. The fifteen that are permitted are shown in Table 30. Since one microstate has $M_L = 2$ ($M_S = 0$) there must be a D-term and this necessitates components of like multiplicity ($S = 0$) that is with $M_L = 1$, 0, -1 and -2; these five microstates will all have the same energy in the absence

Table 30

	$M_L = +2$	$M_L = +1$	$M_L = 0$	$M_L = -1$	$M_L = -2$
m_l values for the two electrons	$(+1, +1)$	$(+1, 0)$	$(+1, -1)$ $(0, 0)$	$(-1, 0)$	$(-1, -1)$
$M_S = +1$		$+1\alpha, 0\alpha$	$+1\alpha, -1\alpha$	$-1\alpha, 0\alpha$	
$M_S = 0$	$+1\alpha, +1\beta$	$\begin{cases} +1\alpha, 0\beta \\ +1\beta, 0\alpha \end{cases}$	$+1\alpha, -1\beta$ $0\alpha, 0\beta$	$-1\alpha, 0\beta$	$-1\alpha, -1\beta$
$M_S = -1$		$+1\beta, 0\beta$	$+1\beta, -1\beta$	$-1\beta, 0\beta$	

$m_s = +\frac{1}{2} = \alpha; m_s = -\frac{1}{2} = \beta$

of a magnetic field. Furthermore, the microstate $M_L = +1, M_S = +1$ must be a component of a ^3P term, the other eight components have all remaining possible values of $M_L = \pm 1, 0$ and $M_S = \pm 1, 0$. The allocation of microstates to ^1D and

^3P terms leaves one, $M_L = 0$, $M_S = 0$ over; this is a ^1S term. To summarize, the three possible terms for the $1s^2 2s^2 2p^2$ configuration of the carbon atom are ^1S, ^1D and ^3P. The corresponding energy levels are $^1S_0(1)$, $^1D_2(5)$, $^3P_2(5)$, $^3P_1(3)$, $^3P_0(1)$; the figures in brackets indicate the number of degenerate microstates in each level $(2J+1)$. In the ^3P term the electron-spin wave function is symmetric to electron interchange; thus both electrons cannot be in the same orbital since the over-all electronic wave function must be antisymmetric. Electron repulsion in this term will therefore be less than that for the singlet terms and the ground state of carbon will be ^3P (see problem 6.1).

Had the two electrons been placed one in each of two p-orbitals of different principal quantum number then none of the Pauli restrictions would have held and for each electron there would have been six different possible m_l, m_s combinations so that for the two electrons there could have been thirty-six possible microstates.

When there are more than two electrons the number of possible microstates increases rapidly. Problems arise only when electrons are to be placed in orbitals of the same n and l. If x electrons are to be disposed $(x < 4l+2)$ then the number of possible microstates is $\{(4l+2)!/(4l+2-x)!\} \div x!$ (divided by $x!$ because electrons are indistinguishable). Thus for an np^3 configuration there will be twenty possible microstates. A simple way of determining the atomic terms that will arise is to seek out the maximum possible values of M_L and M_S, see what terms these require, and then subtract that number of microstates from the total. Thus the highest value of M_L in np^3 will be $(+1\alpha, +1\beta, 0\alpha$, say) 2, also the associated M_S is $+\frac{1}{2}$, therefore a ^2D term which has ten components. The highest possible M_S is $\frac{3}{2}$ but this must be $(+1\alpha, 0\alpha, -1\alpha)$, that is $M_L = 0$, ^4S term (four components) which leaves six possible microstates ^2P. In a similar way the terms for np^x and nd^x may be found; they are listed in Table 31.

Table 31

Electronic configuration	Terms	Number of possible microstates
p^0, p^6	^1S	1
p^1, p^5	^2P	6
p^2, p^4	$^1S\,^3P\,^1D$	15
p^3	$^4S\,^2P\,^2D$	20
d^0, d^{10}	^1S	1
d^1, d^9	^2D	10
d^2, d^8	$^1S\,^3P\,^1D\,^3F\,^1G$	45
d^3, d^7	$^2P\,^4P\,^2D\,^2D\,^2F\,^4F\,^2G\,^2H$	120
d^4, d^6	$^1S^1S^3P^3P^1P^1D^1D^3D^5D^1F^3F^3F^1G^1G^3G^3G^1H^1I$	210
d^5	$^2S\,^6S\,^2P\,^4P\,^2D\,^2D\,^4D\,^2F\,^2F\,^4F\,^2G\,^2G\,^4G\,^2H\,^2I$	252

When subject to an electric field that is not spherically symmetric many of the degeneracies between various electronic microstates are removed. The nature of the splitting of the atomic terms will depend on the symmetry of the perturbing field and the way in which the angular momentum transforms under the field's symmetry operations. The symmetry properties of the atomic orbitals, s, p, d, etc., depend upon their angular moments and since it is possible to represent the real parts of these orbitals pictorially their transformation properties may be determined visually. Thus the character for a rotation through θ radians about the z-axis $(C_{2\pi/\theta})$ upon the three p-orbitals p_x, p_y and p_z is $1 + 2\cos\theta$. For real orbitals the character for p_x and p_y is derived from a second-order matrix; however if linear combinations involving imaginary coefficients are taken a reducible matrix may be obtained.

$$C_{2\pi/\theta}(p_x) = p_x\cos\theta + p_y\sin\theta,$$

$$C_{2\pi/\theta}(p_y) = p_x(-\sin\theta) + p_y\cos\theta.$$

Thus
$$\begin{aligned} C_{2\pi/\theta}(p_x + ip_y) &= p_x(\cos\theta - i\sin\theta) + p_y(\sin\theta + i\cos\theta) \\ &= p_x(\cos\theta - i\sin\theta) + ip_y(\cos\theta - i\sin\theta) \\ &= (p_x + ip_y)e^{-i\theta}. \end{aligned}$$

Also
$$\begin{aligned} C_{2\pi/\theta}(p_x - ip_y) &= p_x(\cos\theta + i\sin\theta) + p_y(\sin\theta - i\cos\theta) \\ &= p_x(\cos\theta + i\sin\theta) - ip_y(\cos\theta + i\sin\theta) \\ &= (p_x - ip_y)e^{+i\theta}. \end{aligned}$$

Hence
$$C_{2\pi/\theta}\begin{bmatrix} p_x + ip_y \\ p_x - ip_y \end{bmatrix} = \begin{bmatrix} e^{-i\theta} & 0 \\ 0 & e^{+i\theta} \end{bmatrix}\begin{bmatrix} p_x + ip_y \\ p_x - ip_y \end{bmatrix}.$$

Thus the characters are

$$\chi C_{2\pi/\theta}(p_x + ip_y) = e^{-i\theta}$$

and $\quad \chi C_{2\pi/\theta}(p_x - ip_y) = e^{+i\theta}.$

The values of m_l for the d-orbitals are ± 2, ± 1 and 0. Let us now consider the rotation through an arbitrary angle on the corresponding family of five real orbitals. A degenerate pair are formed by $d_{x^2-y^2}$ and d_{xy} with an angle of orthogonality of $45°$ ($\delta = 2$). Also d_{xz} and d_{yz} are degenerate, the angle of orthogonality being $90°$ ($\delta = 1$). Under this rotation d_{z^2} is unvariant ($\delta = 0$). As with p-orbitals linear combinations with imaginary coefficients can yield reducible matrices.

Thus $\qquad \chi C_{2\pi/\theta}(d_{xz} \pm id_{yz}) = e^{-i\theta}$

and also $\quad \chi C_{2\pi/\theta}(d_{x^2-y^2} \pm id_{xy}) = e^{\mp i2\theta}.$

Note that $\quad \chi C_{2\pi/\theta}(p_z \text{ or } d_{z^2}) = e^{i0\theta}(= +1).$

For the family of five d-orbitals the rotation through θ may be summarized in a single matrix,

$$C_{2\pi/\theta}\begin{bmatrix} d_{x^2-y^2}+id_{xy} \\ d_{xz}+id_{yz} \\ d_{z^2} \\ d_{xz}-id_{yz} \\ d_{x^2-y^2}-id_{xy} \end{bmatrix} = \begin{bmatrix} e^{-i2\theta} & 0 & 0 & 0 & 0 \\ 0 & e^{-i1\theta} & 0 & 0 & 0 \\ 0 & 0 & e^{i0\theta} & 0 & 0 \\ 0 & 0 & 0 & e^{+i1\theta} & 0 \\ 0 & 0 & 0 & 0 & e^{+i2\theta} \end{bmatrix}$$

This may be generalized for any set of spherical harmonic functions of angular quantum number l, thus

$$C_{2\pi/\theta}\begin{bmatrix} \phi_l \\ \phi_{l-1} \\ \vdots \\ \phi_0 \\ \vdots \\ \phi_{-l} \end{bmatrix} = \begin{bmatrix} e^{-il\theta} & 0 & \cdots & 0 & 0 \\ 0 & e^{-i(l-1)\theta} & \cdots & 0 & 0 \\ \vdots & \vdots & & \vdots & \vdots \\ 0 & 0 & \cdots & e^{i0\theta} & 0 \\ \vdots & \vdots & & \vdots & \vdots \\ 0 & 0 & \cdots & 0 & e^{+il\theta} \end{bmatrix}\begin{bmatrix} \phi_l \\ \phi_{l-1} \\ \vdots \\ \phi_0 \\ \vdots \\ \phi_{-l} \end{bmatrix}$$

The over-all character is

$$\chi = e^{-il\theta}(e^{i2l\theta}+\cdots+e^{i0\theta})$$

$$= e^{-il\theta}\sum_{r=0}^{2l} e^{i\theta r}$$

$$= e^{-il\theta}\left[\frac{e^{i(2l+1)\theta}-1}{e^{i\theta}-1}\right]$$

$$= \frac{e^{i(l+1)\theta}-e^{-il\theta}}{e^{i\theta}-1}$$

$$= \frac{(e^{i(l+\frac{1}{2})\theta}-e^{-i(l+\frac{1}{2})\theta})e^{\frac{1}{2}i\theta}}{e^{i\theta}-1}$$

$$= \frac{2i\sin(l+\frac{1}{2})\theta}{e^{\frac{1}{2}i\theta}-e^{-\frac{1}{2}i\theta}}$$

$$= \frac{2i\sin(l+\frac{1}{2})\theta}{2i\sin\frac{1}{2}\theta}$$

$$= \frac{\sin(l+\frac{1}{2})\theta}{\sin\frac{1}{2}\theta}. \qquad\qquad \textbf{6.2}$$

This is the character for the $2l+1$ orbitals associated with the quantum number l. Since L is a quantum number for atomic states analogous to l, equation **6.2** may also be used to give the character for any rotation upon any atomic term of known L. Table 32 gives the characters for the operations C_2, C_3, C_4, C_5 and C_6 upon various values of L.

Table 32

L	$C_2(\theta = 180°)$	$C_3(\theta = 120°)$	$C_4(\theta = 90°)$	$C_5(\theta = 72°)$	$C_6(\theta = 60°)$
$S:0$	$+1$	$+1$	$+1$	$+1$	$+1$
$P:1$	-1	0	$+1$	$+2\cos36°$	$+2$
$D:2$	$+1$	-1	-1	0	$+1$
$F:3$	-1	$+1$	-1	$-2\cos36°$	-1
$G:4$	$+1$	0	$+1$	-1	-2
$H:5$	-1	-1	$+1$	$+1$	-1
$I:6$	$+1$	$+1$	-1	$+2\cos36°$	$+1$

The effect of the octahedral field, ignoring spin–orbit coupling, will be to reduce the $(2L+1)$-fold degeneracy of the atomic term characterized by the orbital angular-momentum quantum number L. The representations of the resulting subterms are given, for various values of L, in Table 33, using characters from Table 32. (For simplicity only the operations of O-group are considered.)

Table 33

L	E	C_3	$C_2(C_4^2)$	C_2'	C_4	Γs
$S:0$	1	$+1$	$+1$	$+1$	$+1$	A_1
$P:1$	3	0	-1	-1	$+1$	T_1
$D:2$	5	-1	$+1$	$+1$	-1	$E+T_2$
$F:3$	7	$+1$	-1	-1	-1	$A_2+T_1+T_2$
$G:4$	9	0	$+1$	$+1$	$+1$	$A_1+E+T_1+T_2$
$H:5$	11	-1	-1	-1	$+1$	$E+T_1+2T_2$
$I:6$	13	$+1$	$+1$	$+1$	-1	$A_1+A_2+E+2T_1+T_2$

Very often in coordination compounds the ligands' symmetry about the central atom is only approximately octahedral; the field may be regarded as basically octahedral with a weaker field of lower symmetry superimposed upon it. This will have the effect of causing further splitting of the representations which were degenerate under O_h. For example the complex cation $Ti(H_2O)_6^{3+}$ has but one d-electron. The term for the ground electronic state will be ^2D and in the presence of the octahedral field due to the water molecules this will be split into (using O_h symbolism) $^2T_{2g}$ and ^2E_g. The colour of the titanous ion is thought to be due to a transition from the former to the latter state ($v_{max} = 2030$ mm^{-1}). Furthermore the arrangement of water molecules about Ti^{3+} is subject to a slight trigonal distortion, the threefold axis of distortion being coincident with one of the three-fold axes of O_h. To find out what effect this will have on T_{2g} and E_g allow functions characterized by each of the irreducible representations of O_h to be subject to the symmetry operations of D_{3d}. More simply the same sort of information may be obtained by constructing a reduction table from O to D_3 (Table 34). (For other examples involving a reduction in symmetry, see p. 153.)

Table 34

Representations in O	Operations D_3			Representations in D_3
	E	$2C_3$	$3C_2$	
A_1	1+	1+	1	A_1
A_2	1+	1	−1	A_2
E	2	−1	0	E
T_1	3	0	−1	A_2+E
T_2	3	0	+1	A_1+E

Thus the distortion has no effect upon ^2E_g but splits the lower $^2T_{2g}$ subterm into $^2A_{2g}$ and ^2E_g (Figure 77).

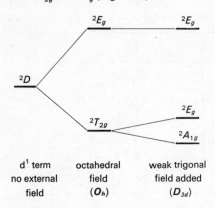

$$^2E_g \qquad ^2E_g$$

$$^2D$$

$$^2T_{2g}$$

$$^2E_g$$

$$^2A_{1g}$$

d^1 term no external field	octahedral field (O_h)	weak trigonal field added (D_{3d})

Figure 77   'Splitting' of a ^2D term in an octahedral field and the subsequent 'splitting' of the $^2T_{2g}$ subterm by an additional trigonal field

6.3.3   Strong fields

So far it has been assumed that the crystal field is quite strong enough to break up the coupling between L and S in the atom (or ion) but not strong enough to break up the couplings between m_ls or m_ss that produced L and S respectively. Such a field is called a *weak field*; a *strong field* is one that *is* intense enough to destroy the coupling between the m_ls or the m_ss so that L and S are no longer good quantum numbers.

If the strong field has octahedral symmetry the five degenerate d-orbitals of the free gaseous atom (or ion) will be split into a pair of e_g orbitals and a trio of t_{2g} orbitals. These two sets of orbitals will be widely separated in energy (Δ in Figure 56, p. 128) and the electron population of each energy level must be considered individually. It will therefore be necessary to consider the way in which electron configurations such as t_{2g}^x and e_g^y (x and y, numbers of electrons) are affected by symmetry operations in much the same way as we have considered the effect of such operations upon atomic configurations. The problem is somewhat simplified since sets of t_{2g} orbitals will have symmetry properties similar to p (t_{1u}) orbitals.

Consider the effect of an octahedral field upon various possible t_{2g}^x:

t_{2g}^1 and t_{2g}^5 will transform as t_{2g}, therefore each gives a $^2T_{2g}$ term.

For t_{2g}^2 and t_{2g}^4 the direct product of t_{2g}^2 may be reduced to $a_{1g}+t_{1g}+e_g+t_{2g}$. This is the same combination of irreducible representations as arises from the atomic configuration p^2 under the influence of a field of O_h symmetry. The terms $^1S+{}^3P+{}^1D$ become $^1A_{1g}+{}^3T_{1g}+{}^1E_g+{}^1T_{2g}$. These are therefore also the terms that arise from t_{2g}^2 and t_{2g}^4.

The configuration t_{2g}^3 is in some ways analogous to p^3. The latter yields terms ^4S, ^2P, ^2D. If we attempt to build up the t_{2g}^3 terms from those that arise from t_{2g}^2 under an O_h field then quartet terms can only come from the $^3T_{1g}$ term in combination with the third electron in a t_{2g} orbital, that is in a $^2T_{2g}$ state: the product representations upon reduction yield $^{2\,or\,4}(A_{2g}+E_g+T_{1g}+T_{2g})$. Clearly a quartet term can only be associated with one electron in each of the three t_{2g} orbitals. Thus the only possible quartet term is $^4A_{2g}$. The other terms follow more simply from the p^3 analogy: ^2P and ^2D become $^2T_{1g}$ and $^2E_g+{}^2T_{2g}$ (gerade because actually we are dealing with d-orbitals not p-orbitals). To summarize, t_{2g}^3 gives rise to $^4A_{2g}+{}^2E_g+{}^2T_{1g}+{}^2T_{2g}$.

In the degenerate pair of e_g orbitals, e_g^1 and e_g^3 give rise to the doublet term ^2E_g, but for e_g^2 both singlet and triplet terms are possible. The orbital functions from e_g^2 under an octahedral field are $A_{1g}+A_{2g}+E_g$.

If the two orthogonal orbitals that comprise e_g are ϕ_1 and ϕ_2 then it is possible to envisage four possible allocations of two electrons to them:

$$\psi_1 = \phi_1(1)\phi_1(2),$$

$$\psi_2 = \phi_2(1)\phi_2(2),$$

$$\psi_3 = \phi_1(1)\phi_2(2)+\phi_1(2)\phi_2(1),$$

$$\psi_4 = \phi_1(1)\phi_2(2)-\phi_1(2)\phi_2(1).$$

These functions are not normalized; it is their symmetry properties that are important. The wave functions ψ_1, ψ_2 and ψ_3 are all symmetric with respect to electron interchange. These functions must therefore be associated with antisymmetric spin functions (i.e. singlet). However ψ_4 is an antisymmetric orbital function and so must be combined with a symmetric (triplet) spin function. Clearly ψ_4 can only correspond to A_{1g} or A_{2g}.

The functions ϕ_1 and ϕ_2 will correspond to the orbitals d_{z^2} and $d_{x^2-y^2}$ whose symmetry properties under the operations of the point group O are (O has the essential features of O_h for this argument and is simpler to use):

$$E(\phi_1) = +1(\phi_1) \qquad\qquad E(\phi_2) = +1(\phi_2),$$

$$C_3(\phi_1) = (\cos 120°)\phi_1 + (\sin 120°)\phi_2 \qquad C_3(\phi_2) = (-\sin 120°)\phi_1 + (\cos 120°)\phi_2,$$

$$C_2(\phi_1) = +1(\phi_1) \qquad\qquad C_2(\phi_2) = +1(\phi_2),$$

$$C_2'(\phi_1) = +1(\phi_1) \qquad\qquad C_2'(\phi_2) = -1(\phi_2),$$

$$C_4(\phi_1) = +1(\phi_1) \qquad\qquad C_4(\phi_2) = -1(\phi_2).$$

We can now evaluate the effect of O-group symmetry operators on ψ_4:

$$E(\psi_4) = +1(\psi_4),$$

$$C_3(\psi_4) = [(\cos 120°)\phi_1(1) + (\sin 120°)\phi_2(1)][(-\sin 120°)\phi_1(2) + (\cos 120°)\phi_2(2)] -$$
$$- [(\cos 120°)\phi_1(2) + (\sin 120°)\phi_2(2)][(-\sin 120°)\phi_1(1) + (\cos 120°)\phi_2(1)]$$

$$= (\cos^2 120°)\phi_1(1)\phi_2(2) - (\sin^2 120°)\phi_1(2)\phi_2(1) + (\sin^2 120°)\phi_1(1)\phi_2(2) -$$
$$- (\cos^2 120°)\phi_1(2)\phi_2(1)$$

$$= \phi_1(1)\phi_2(2) - \phi_1(2)\phi_2(1)$$

$$= +1(\psi_4),$$

$$C_2(\psi_4) = +1(\psi_4),$$

$$C_2'(\psi_4) = (+1)\phi_1(1)(-1)\phi_2(2) - (+1)\phi_1(2)(-1)\phi_2(1)$$

$$= -1(\phi_4),$$

$$C_4(\psi_4) = -1(\psi_4).$$

Thus $\Gamma(\psi_4) = A_{2g}$. The terms from e_g^2 are therefore $^1A_{1g} + {}^3A_{2g} + {}^1E_g$.

Another method by which the terms that arise from an octahedral field acting upon t_{2g}^x or e_g^y configurations can be evaluated involves a reduction of symmetry. Let us consider distortions of the O point group, this will usually contain sufficient information and will be much less complicated than using the O_h group. The reduction to D_3 symmetry has been given in Table 34: Table 35 is the corresponding correlation with D_4.

Thus the configuration e^2 in O would become, upon distortion, either a_1^2 or b_1^2 or $a_1^1 b_1^1$. These would give rise to terms $^1A_1$, $^1A_1$ and $^1B_1$ and $^3B_1$. The first two must be singlets being electron pairs in single orbitals whilst $a_1^1 b_1^1$ can give rise to either singlet or triplet terms. Now under O symmetry e^2 gives rise to terms $^xA_1 + {}^yA_2 + {}^zE$; these terms become upon reduction of symmetry

$$^xA_1 + {}^yB_1 + {}^zA_1 + {}^zB_1.$$

Correlation with the terms given above shows that $x = 1$, $y = 3$, $z = 1$ is the only possible solution.

Table 35

$O \longrightarrow$ i.r. in O	E	C_2	C_4	C_2	C_2'	
$D_4 \to$	E	C_2	C_4	C_2'	C_2''	i.r.s in D_4
A_1	1	1	1	1	1	A_1
A_2	1	1	-1	1	-1	B_1
E	2	2	0	2	0	$A_1 + B_1$
T_1	3	-1	1	-1	-1	$E + A_2$
T_2	3	-1	-1	-1	1	$E + B_2$

The same method can be used for t_{2g}^2 (see Table 36).

Table 36

Terms		Configurations		Possible terms
D_3	O	O	D_3	D_3
$^aA_1$	$^aA_1$		e^2	$^3A_2, {}^1A_1, {}^1E$
^bE	^bE	t_2^2	$e^1a_1^1$	$^1E, {}^3E$
$^cA_2 + {}^cE$	$^cT_1$		a_1^2	$^1A_1$
$^dA_1 + {}^dE$	$^dT_2$			

Correlation of the columns of the far left and far right shows that $c = 3$ and that $a = b = d = 1$.

Similarly for t_{2g}^3 using reduction of O to D_3 we have

$$t_2^3 \begin{cases} e^3 & \to {}^2E \\ e^2a_1^1 \to \begin{cases} {}^3A_2 \, {}^2A_1 = {}^4A_2, \, {}^2A_2 \\ {}^1A_1 \, {}^2A_1 = {}^2A_1 \\ {}^1E \, {}^2A_1 = {}^2E \end{cases} \\ e^1a_1^2 \to {}^2E \end{cases}$$

The allocation of three electrons to three orbitals would give rise to terms ^4S, ^2P, ^2D in the absence of any crystal field. Correlation of terms of like multiplicity shows that in O the terms are $^4A_2, {}^2T_1, {}^2T_2$ and ^2E, since when symmetry is further reduced, the terms are as given above.

The terms that group all possible arrangements of electrons in strong-field configurations may be found by simply evaluating the direct products of the terms for the t_{2g} and e_g populations. The number of microstates in the over-all term (i.e. its degeneracy) will be the product of the degeneracies of the t_{2g} and e_g terms. These may be found from the product of the multiplicity and orbital degeneracy of each individual term. Thus the degeneracy of the configuration $t_{2g}^3 e_g^2$ is 120. The terms for t_{2g}^3 are $^4A_{2g} + {}^2E_g + {}^2T_{1g} + {}^2T_{2g}$ so that the number of microstates in t_{2g}^3 is $4 \times 1 + 2 \times 2 + 2 \times 3 + 2 \times 3 = 20$, similarly for e_g^2 $({}^1A_{1g} + {}^1E_g + {}^3A_{2g})$, $1 \times 1 + 1 \times 2 + 3 \times 1 = 6$.

A most important difference between weak-field and strong-field complexes is that the ground state of the former has maximum multiplicity for d^n (like a free atom) while the latter has maximum multiplicity for t_{2g}^n or $t_{2g}^6 e_g^n$ where n is a number of electrons. Thus ground states for d^4, d^5, d^6 and d^7 will be drawn from configurations $t_{2g}^3 e_g^1$, $t_{2g}^3 e_g^2$, $t_{2g}^4 e_g^2$ and $t_{2g}^5 e_g^2$ for weak fields but from $t_{2g}^4 e_g^0$, $t_{2g}^5 e_g^0$, $t_{2g}^6 e_g^0$ and $t_{2g}^6 e_g^1$ for strong fields. Energy-level diagrams which show how the various terms are affected by increasing field have been devised by Orgel ('Spectra of transition-metal complexes', *Journal of Chemical Physics*, vol. 23 (1955), no. 6, pp. 1004–14; 'Band widths in the spectra of manganous and other transition-metal ions', *Journal of Chemical Physics*, vol. 23 (1955), no. 10, pp. 1824–6) and by Tanabe and Sugano ('On the absorption spectra of complex ions. I and II', *Journal of the Physical Society of Japan*, vol. 9 (1954), no. 5, pp. 753–66 and 766–79).

Whilst it is not the purpose of this book to describe the calculations necessary for the construction of these diagrams, it is possible to obtain some qualitative conclusions for simple cases by correlating like terms in weak and strong fields.

Consider a transition metal with two d-electrons, V^{3+} for example; the ground state of the gaseous ion will be a triplet, ^3F. There is also an excited state of the same multiplicity to which transition will be possible without change of electron spin (^3P). There are also three singlet terms ^1S, ^1D and ^1G but since none is the ground state electronic transition between these terms or to these terms from ^3F will be very weak indeed.

The ^3F term will be split by a weak octahedral field into $^3A_{2g}$, $^3T_{1g}$, $^3T_{2g}$ but ^3P is unaffected, yielding $^3T_{1g}$ (Table 33). In the strong field case the starting point must be the electron populations in the t_{2g} and e_g orbitals. The most stable configuration will be t_{2g}^2 which is split by the field into $^1A_{1g}$, $^3T_{1g}$, ^1E_g, $^1T_{2g}$. The first excited configuration is $t_{2g} e_g$ which yields terms $^1T_{1g} + {}^3T_{1g} + {}^1T_{2g} + {}^3T_{2g}$ whilst the terms from the second excited configuration e_g^2 are $^1A_{1g} + {}^3A_{2g} + {}^1E_g$. Figure 78 shows in a qualitative way how the various triplet terms may be correlated. The complete energy-level diagram which includes the singlet is of course more complicated; unfortunately the diagrams for the most interesting cases where the ground state changes multiplicity are very much more complicated and cannot be dealt with here.

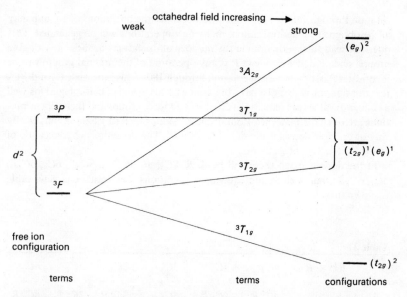

octahedral field increasing →

weak   strong

$\overline{(e_g)^2}$

$^3A_{2g}$

^3P $^3T_{1g}$

d^2 $\left.\begin{array}{c}\\\\\end{array}\right\}(t_{2g})^1(e_g)^1$

$^3T_{2g}$

^3F

$^3T_{1g}$

free ion
configuration

$\underline{\quad}(t_{2g})^2$

terms   terms   configurations

Figure 78   Qualitative energy-level correlation diagram for triplet states derived
from a d^2 configuration (e.g. V^{3+}) under the influence of an increasing
octahedral ligand field

6.4   Spin–orbit coupling – double groups

The angular momentum of orbitals is governed by integral quantum numbers
and the general character for a rotational symmetry operation upon a group of
quantum numbers is given by equation **6.2**. The same equation may also be used
for atomic-orbital angular-momentum quantum numbers L and for the total
angular-momentum quantum numbers J. However, since electrons have spin
quantum numbers of $\pm\frac{1}{2}$, S, and J too, may have half-integral as well as integral
values. The introduction of half-integral quantum numbers into equation **6.2**
raises a novel problem and necessitates the formulation of double groups.

Let us consider the simplest case where the quantum number (be it S or J)
is $\frac{1}{2}$.

$$\chi C_{2\pi/\theta} = \frac{\sin(\frac{1}{2}+\frac{1}{2})\theta}{\sin\frac{1}{2}\theta} = \frac{2\sin\frac{1}{2}\theta\cos\frac{1}{2}\theta}{\sin\frac{1}{2}\theta} = 2\cos\frac{1}{2}\theta$$

The simplest operation of all, E, is equivalent to a rotation through 360°. The
above character then takes on the value of $2\cos 180°$, i.e. -2. Only rotation
through 720° (and its multiples) will yield $\chi = +2$. In order to handle these
representations involving electron spin it is necessary to extend the simple point-
group tables used up to now. One must imagine the rather extraordinary situation in which only a rotation through 720° is the identity operator and that

rotation through $360°$ is a new operation R. Since the operations of a group may all be obtained by multiplication, many new operations will be generated. This does not lead to exactly double the number of operations in these new double groups since P and RP (where P is any operation of the original group) will be identical if P is, or contains, a rotation through $180°$. This is because C_2 and RC_2 (i.e. rotation through $540°$) have the same character for all half-integral (as well as integral) values of l (equation **6.2**), $\chi = 0$. Double groups and their representations are often distinguished from ordinary groups by a single dash, thus O'. By way of example consider the double group O'. The operations characteristic of O are E, $8C_3$, $3C_2$, $6C_4$ and $6C_2'$.

In the double group there will be E, R, $8C_3$, $8C_3R$, $3C_2$, $3C_2R$, $6C_4$, $6C_4R$, $6C_2'$, $6C_2'R$; that is eight non-equivalent operations. The actual character table is shown in Table 37.

Table 37

O'	E	R	$4C_3$ $+$ $4C_3^2 R$	$4C_3^2$ $+$ $4C_3 R$	$3C_2$ $+$ $3C_2 R$	$3C_4$ $+$ $3C_4^3 R$	$3C_4^3$ $+$ $3C_4 R$	$6C_2'$ $+$ $6C_2' R$
$\Gamma_1 A_1'$	1	1	1	1	1	1	1	1
$\Gamma_2 A_2'$	1	1	1	1	1	-1	-1	-1
$\Gamma_3 E_1'$	2	2	-1	-1	2	0	0	0
$\Gamma_4 T_1'$	3	3	0	0	-1	1	1	-1
$\Gamma_5 T_2'$	3	3	0	0	-1	-1	-1	1
$\Gamma_6 E_2'$	2	-2	1	-1	0	$\sqrt{2}$	$-\sqrt{2}$	0
$\Gamma_7 E_3'$	2	-2	1	-1	0	$-\sqrt{2}$	$\sqrt{2}$	0
$\Gamma_8 G'$	4	-4	-1	1	0	0	0	0

Table 37 may now be used to determine the representations, under octahedral symmetry, of species whose angular-momentum quantum numbers are half-integral. These are summarized in Table 38 (the results for integral values, taken from Table 33, are included for convenience).

Consider by way of example the case of one d-electron. In the gaseous ion the electronic term is $^2D(L = 2, S = \frac{1}{2})$. Spin–orbit coupling gives two possible values of J, $2+\frac{1}{2} = \frac{5}{2}$ and $2-\frac{1}{2} = \frac{3}{2}$; the two energy levels may be symbolized $^2D_{\frac{5}{2}}$ (6), $^2D_{\frac{3}{2}}$ (4). The figures in parentheses give the degeneracies of the levels $(2J+1)$, i.e. the number of contributory microstates. If now an octahedral field is introduced some of the degeneracy of $^2D_{\frac{5}{2}}$ is removed, two groups of states are formed, Γ_7 and Γ_8. $^2D_{\frac{3}{2}}$ is unaffected by the field, Γ_8. The two Γ_8 levels, having the same representation, may interact: the stronger the spin–orbit coupling the more they will do so.

Table 38

Angular-momentum quantum numbers	E	R	$8C_3$	$8C_3R$	$6C_2$	$6C_4$	$6C_4R$	$12C_2'$	Γs
0	1								Γ_1
$\frac{1}{2}$	2	-2	1	-1	0	$\sqrt{2}$	$-\sqrt{2}$	0	Γ_6
1	3								Γ_4
$1\frac{1}{2}$	4	-4	-1	1	0	0	0	0	Γ_8
2	5								$\Gamma_3+\Gamma_5$
$2\frac{1}{2}$	6	-6	0	0	0	$-\sqrt{2}$	$\sqrt{2}$	0	$\Gamma_7+\Gamma_8$
3	7								$\Gamma_2+\Gamma_4+\Gamma_5$
$3\frac{1}{2}$	8	-8	1	-1	0	0	0	0	$\Gamma_6+\Gamma_7+\Gamma_8$
4	9								$\Gamma_1+\Gamma_3+\Gamma_4+\Gamma_5$
$4\frac{1}{2}$	10	-10	-1	1	0	$\sqrt{2}$	$-\sqrt{2}$	0	$\Gamma_6+2\Gamma_8$
5	11								$\Gamma_3+\Gamma_4+2\Gamma_5$
$5\frac{1}{2}$	12	-12	0	0	0	0	0	0	$\Gamma_6+\Gamma_7+2\Gamma_8$
6	13								$\Gamma_1+\Gamma_2+\Gamma_3+$ $+2\Gamma_4+\Gamma_5$
$6\frac{1}{2}$	14	-14	1	-1	0	$-\sqrt{2}$	$\sqrt{2}$	0	$\Gamma_6+2\Gamma_7+2\Gamma_8$

Note that the column E gives the degeneracy.

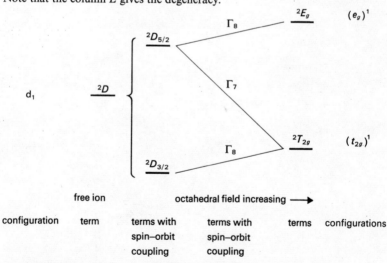

spin–orbit coupling effects, decreasing relative to ligand field influence ⟶

Figure 79　Relative effects of spin–orbit coupling and an octahedral ligand field upon a d^1 configuration

In the strong-field case the possible terms for a single d-electron are $^2T_{2g}$ and ^2E_g. The representations for the energy levels that result when spin–orbit coupling is included may be found from the direct product $\Gamma(S)\Gamma(\text{orbital})$. Thus for $^2T_{2g}$, $S = \frac{1}{2}$ so the product is $\Gamma(S = \frac{1}{2})\Gamma(T_{2g}) = \Gamma_6\Gamma_5 = \Gamma_7 + \Gamma_8$; similarly for ^2E_g the product is $\Gamma(S = \frac{1}{2})\Gamma(E_g) = \Gamma_6\Gamma_3 = \Gamma_8$ (O' has been used rather than O'_h and so the gs have been ignored). These effects are shown diagrammatically in Figure 79. On the left there is no octahedral field but there is spin–orbit coupling. On moving to the right the former is intensified and the latter reduced until the spin–orbit effect is zero. It is important to realize that the diagram shows *relative* effects. The *absolute* effect of either spin–orbit coupling or the octahedral field may of course be either large or small. Large spin–orbit effects lead to j–j coupling whereas small effects can be dealt with by L–S coupling.

Problem 6

6.1   The possible spin functions for two electrons have already been considered (Equations 1.4 – 1.7).

Write down the possible orbital functions for two electrons allocated to a set of $2p$ orbitals and note the symmetry of each function. Thus deduce the number of possible microstates.

Chapter 7
Molecular Vibrations
and Crystal Symmetry

In this chapter we shall look rather briefly at some other applications of symmetry ideas to molecular problems. Molecular vibrations and selection rules for infrared and Raman spectra will be examined first. In the second part the role of symmetry in crystal structures will be outlined.

7.1 Infrared and Raman spectroscopy

The vibrations of a molecule can most simply be expressed in terms of simple harmonic oscillation of the molecule as a whole in 'normal coordinates' (Q). Each normal coordinate would in principle involve the displacement of every atom in the molecule in a particular direction (different for each atom) for a specified distance. The symmetry properties of these coordinates can be found by considering the potential motion of each atom in three mutually perpendicular directions (e.g. parallel to x-, y- and z-axes of the molecule). Place three arrows to represent $\mathbf{x}$, $\mathbf{y}$ and $\mathbf{z}$ at each atom and determine the representation for the family of $3N$ arrows (an N-atom molecule). Each arrow vector will transform like the corresponding p-orbital. This representation will not only include representations for all possible modes of vibration of the molecule, but also representations for its rotation and translation in space. The translation and rotation of the whole molecule can each be completely described using six of these coordinates so there will be $3N - 6$ vibrational Q-coordinates. Should the molecule have a C_∞ axis of symmetry, rotations about the x- and y-axes become equivalent and any rotation can be completely described using only two coordinates leaving $3N - 5$ for vibration. The representations for the coordinates of translation and rotation must be determined separately and then subtracted from the representation of the $3N$ atomic vectors to leave the representation of the normal vibrational coordinates of the molecule. This representation may be reduced in the usual way yielding the symmetries of the normal vibrations as irreducible representations of the symmetry group of the molecule. The determination of the exact nature of the normal vibrations is much more complex but it is usually possible to suggest pictorially the basic features of the atomic motions.

7.1.1 Examples

Water (C_{2v}). This molecule has three atoms and all possible motions may be

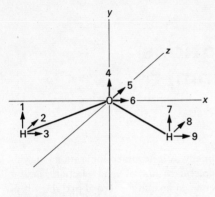

Figure 80   Resolution parallel to Cartesian axes of the potential displacements of the atoms in a water molecule

described using nine coordinates. The small arrows in Figure 80 show how potential displacements may be resolved in directions parallel to Cartesian axes. These arrows may now be subjected to the operations of C_{2v} just as were atomic orbitals on page 88. The matrix for the operation E is surely obvious but by way of illustration the over-all matrices will be written out for C_2, $\sigma_v'(yz)$ and $\sigma_v(xz)$.

C_2

$$
\begin{array}{c}
1 \\ 2 \\ 3 \\ 4 \\ 5 \\ 6 \\ 7 \\ 8 \\ 9
\end{array}
\begin{bmatrix}
0 & 0 & 0 & 0 & 0 & 0 & -1 & 0 & 0 \\
0 & 0 & 0 & 0 & 0 & 0 & 0 & +1 & 0 \\
0 & 0 & 0 & 0 & 0 & 0 & 0 & 0 & -1 \\
0 & 0 & 0 & -1 & 0 & 0 & 0 & 0 & 0 \\
0 & 0 & 0 & 0 & +1 & 0 & 0 & 0 & 0 \\
0 & 0 & 0 & 0 & 0 & -1 & 0 & 0 & 0 \\
-1 & 0 & 0 & 0 & 0 & 0 & 0 & 0 & 0 \\
0 & +1 & 0 & 0 & 0 & 0 & 0 & 0 & 0 \\
0 & 0 & -1 & 0 & 0 & 0 & 0 & 0 & 0
\end{bmatrix}
$$

Thus $\chi = -1$.

$\sigma_v'(yz)$

$$
\begin{array}{c}
1 \\ 2 \\ 3 \\ 4 \\ 5 \\ 6 \\ 7 \\ 8 \\ 9
\end{array}
\begin{bmatrix}
0 & 0 & 0 & 0 & 0 & 0 & +1 & 0 & 0 \\
0 & 0 & 0 & 0 & 0 & 0 & 0 & +1 & 0 \\
0 & 0 & 0 & 0 & 0 & 0 & 0 & 0 & -1 \\
0 & 0 & 0 & +1 & 0 & 0 & 0 & 0 & 0 \\
0 & 0 & 0 & 0 & +1 & 0 & 0 & 0 & 0 \\
0 & 0 & 0 & 0 & 0 & -1 & 0 & 0 & 0 \\
+1 & 0 & 0 & 0 & 0 & 0 & 0 & 0 & 0 \\
0 & +1 & 0 & 0 & 0 & 0 & 0 & 0 & 0 \\
0 & 0 & -1 & 0 & 0 & 0 & 0 & 0 & 0
\end{bmatrix}
$$

Thus $\chi = +1$.

$\sigma_v(xz)$

$$
\begin{array}{c}
1 \\ 2 \\ 3 \\ 4 \\ 5 \\ 6 \\ 7 \\ 8 \\ 9
\end{array}
\begin{bmatrix}
-1 & 0 & 0 & 0 & 0 & 0 & 0 & 0 & 0 \\
0 & +1 & 0 & 0 & 0 & 0 & 0 & 0 & 0 \\
0 & 0 & +1 & 0 & 0 & 0 & 0 & 0 & 0 \\
0 & 0 & 0 & -1 & 0 & 0 & 0 & 0 & 0 \\
0 & 0 & 0 & 0 & +1 & 0 & 0 & 0 & 0 \\
0 & 0 & 0 & 0 & 0 & +1 & 0 & 0 & 0 \\
0 & 0 & 0 & 0 & 0 & 0 & -1 & 0 & 0 \\
0 & 0 & 0 & 0 & 0 & 0 & 0 & +1 & 0 \\
0 & 0 & 0 & 0 & 0 & 0 & 0 & 0 & +1
\end{bmatrix}
$$

Thus $\chi = +3$.

The representation for the nine coordinates is therefore

E	C_2	$\sigma_v(xz)$	$\sigma'_v(yz)$
9	-1	$+3$	$+1$

This includes translations and rotations of the whole molecule. The translations in space may be resolved into translations along the x-, y- and z-axes and will transform accordingly.

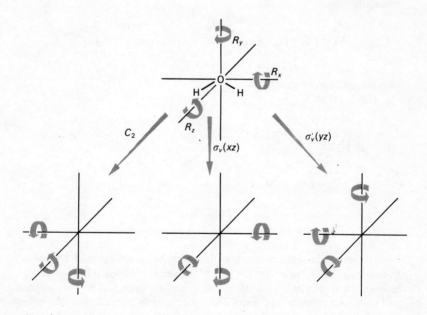

Figure 81   Effect of the symmetry operators, C_2, $\sigma_v(xz)$ and $\sigma_v(yz)$, on the components of the rotation of a water molecule

	E	C_2	$\sigma_v(xz)$	$\sigma'_v(yz)$
Γ_x	1	-1	$+1$	-1
Γ_y	1	-1	-1	$+1$
Γ_z	1	$+1$	$+1$	$+1$

$$\Gamma_{\text{trans}} = \Gamma_x + \Gamma_y + \Gamma_z \rightarrow \Gamma(\mathbf{x}) + \Gamma(\mathbf{y}) + \Gamma(\mathbf{z})$$

3	-1	$+1$	$+1$

The characters for rotations may be found by considering three possible rotations into which any rotation could be resolved, for example about the x-axis (R_x), the y-axis (R_y) and the z-axis (R_z). These three component rotational vectors are illustrated as arrows in Figure 81. If the direction of rotation remains the same, this may be written as $+1$ (R) but if it is reversed, as -1 (R). The representations are

	E	C_2	$\sigma_v(xz)$	$\sigma'_v(yz)$
R_x	1	-1	-1	$+1$
R_y	1	-1	$+1$	-1
R_z	1	$+1$	-1	-1
Γ_{rot}	3	-1	-1	-1

The representations for the three normal vibrations of water may be found by subtraction, thus:

		E	C_1	$\sigma_v(xz)$	$\sigma'_v(yz)$
$\Gamma_{\text{over all}}(3N) = \Gamma_{\text{vib}} + \Gamma_{\text{rot}} + \Gamma_{\text{trans}}$		9	-1	$+3$	$+1$
for Γ_{trans}		3	-1	$+1$	$+1$
and for Γ_{rot}		3	-1	-1	-1

Subtracting the representation for
Γ_{trans} and Γ_{rot} from $\Gamma_{\text{over all}}$:

	Γ_{vib}	3	$+1$	$+3$	$+1$
	Γ_{vib}	$2A_1 + B_1$			

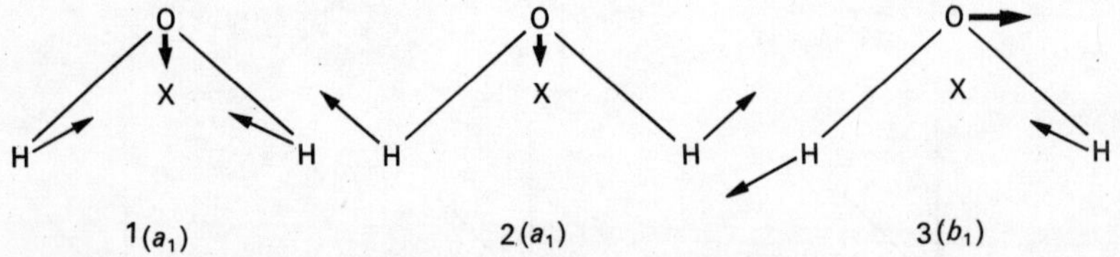

Figure 82   The normal modes of vibration for a water molecule

Vibrations of the water molecule of these symmetries are shown in Figure 82. In each diagram all the arrows are part of one single vibration. The relationship between the arrows of Figures 80 and 82 is analogous to that between atomic and molecular orbitals. The arrows have been drawn in phase but of course, represent only half of the vibration, the equal and opposite restoring motions have been omitted. The arrows are different sizes to indicate different magnitudes of displacement. Since these are only vibrations there must be no movement of the centre of mass (X) of the molecule. Oxygen is sixteen times heavier than hydrogen and so, relative to hydrogen, its displacement from its equilibrium position in any mode of vibration will be small. Also the centre of mass will be very near the oxygen atom so that the vibrations (1 and 3) are almost directed along the O–H bonds. The other mode of vibration (2) is a symmetrical flapping of the O–H bonds.

Ammonia, NH_3 (C_{3v}). All possible motions of the four atoms in ammonia may be described by the use of twelve coordinates. As can be seen from the matrices for water when atoms are moved in space by a symmetry operation the characters for displacement vectors (i.e. the little arrows) which such atoms contribute to the diagonal of the matrix are always zero. As in the case of the atomic orbitals (p. 101) this means that the representation even for quite complex molecules can be written down by inspection. Thus for ammonia,

Operation	Effect	Character
E_1	all arrows unmoved	total, $\chi = 12$
C_3	all arrows on hydrogen moved: $\chi = 0$	
	z-arrow on N unmoved: $\chi = +1$	
	x-, y-arrows on N behave as a degenerate pair:	
	$\chi = -1$	total, $\chi = 0$
σ_v	two hydrogens reflected into each other, all	
	arrows moved: $\chi = 0$	
	other hydrogen, two arrows in σ_v plane: $\chi = 2$	
	one arrow perpendicular to σ_v	
	plane: $\chi = -1$	
	nitrogen, two arrows in σ_v plane: $\chi = 2$	
	one arrow perpendicular to σ_v	
	plane: $\chi = -1$	total, $\chi = 2$

The representations for the translation of the ammonia molecule are those for the x-, y- and z-axes. The representation for the rotations are,

	E	$2C_3$	$3\sigma_v$
R_x, R_y	2	-1	0
R_z	1	1	-1
Thus Γ_{rot}	3	0	-1

The representations for the vibrations can now be found by subtraction:

	E	$2C_3$	$3\sigma_v$
$\Gamma_{over\ all}(3N)$ for NH_3	12	0	2
Γ_{trans}	3	0	1
Γ_{rot}	3	0	-1
Hence Γ_{vib}	6	0	2
$\Gamma_{vib} = 2A_1 + 2E$			

Thus there will be two wholly symmetric vibrations and two degenerate pairs of vibrations.

Solution of the Schrödinger equation for a simple harmonic oscillator shows that the permitted energy levels are

$$E_i(n_i) = (n_i + \tfrac{1}{2})hv_i$$

and that the wave functions have the form

$$\psi_i(n_i) = N_i \exp(-\tfrac{1}{2}\alpha_i^2 Q_i) H_{n_i}(\alpha_i, Q_i) \qquad \textbf{7.1}$$

where v_i is the fundamental frequency of oscillation in the ith coordinate Q_i, n_i is the number of quanta in ith oscillator, $\alpha_i^2 = 4\pi^2 v_i/h$, N_i is a normalizing constant and $H_{n_i}(\alpha_i Q_i)$ is the Hermite polynomial of degree n_i.

The wave function that describes the over-all vibrational state of a molecule will be the product of all the individual oscillations, thus:

$$\Psi_{\text{vib}} = \prod_{i=1}^{x} \psi_i(n_i) \qquad (x = 3N - 6 \text{ or } x = 3N - 5).$$

It will be important to determine the representation of Ψ_{vib}; this will be the direct product of the representations for each oscillator, for example $\Gamma(\psi_i(n_i))$. These representations may be found by examining equation **7.1**. Both N_i and the exponential term will be scalar numbers and therefore invariant under symmetry operations but the Hermite polynomial will contain terms of the type $(Q_i)^{n_i}$ the representation for which will simply be $[\Gamma(Q_i)]^{n_i}$, thus

$$\Gamma(\psi_i(n_i)) = [\Gamma(Q_i)]^{n_i}.$$

When $n_i = 0$ the representation will be wholly symmetric: this will also be the representation for the over-all vibrational wave function of the molecule in its ground vibrational state since for every oscillator n_i will be zero.

Using the Born–Oppenheimer approximation it is possible to express the total wave function for a molecule as a product of electronic, vibrational, rotational, etc., parts, and to consider each type of function independently of the others. Thus it is possible to consider the vibrational wave function in isolation. The probability p of a transition between one vibrational state and another is expressed by an equation similar to equation **6.1**; the symmetry requirements also hold,

$$\Gamma(p) = \Gamma(\text{wholly symmetric}) = \Gamma(\Psi_{\text{I}})\Gamma(\mathscr{P})\Gamma(\Psi_{\text{II}}). \qquad \textbf{7.2}$$

$\mathscr{P}$ = operator effecting transition between states Ψ_{I} and Ψ_{II}.

Transition in the infrared region can be induced by electromagnetic radiation. The electric vector is more important and so

$$\Gamma(\mathscr{P}, \text{infrared}) = \Gamma(\mathscr{P}, \text{i.r.}) = \Gamma(x, y \text{ or } z)$$

just as for ultraviolet transitions considered in Chapter 6. Raman spectra are formed in a more complex way which involves the polarization of the molecule under the influence of electromagnetic radiation – often visible light.

$\Gamma(\mathscr{P}, \text{Raman}) = \Gamma(\mathscr{P}, R) = \Gamma(x^2, y^2, z^2, xy, xz \text{ and } yz)$.

If a molecule possesses a centre of symmetry it is interesting to note that all the representations for $(\mathscr{P}, \text{i.r.})$ are necessarily *ungerade* ($-$ parity) whilst those for $(\mathscr{P}, R)$ are all *gerade* ($+$ parity). Since $\Gamma(\Psi_I)$, if we are considering the molecule in its ground vibrational state, is wholly symmetric, then the product $\Gamma(\mathscr{P})\Gamma(\Psi_{II})$ must also be wholly symmetric, and therefore *gerade*, for the transition not to be forbidden. Thus only excitations to states of *ungerade* symmetry stand a chance of being observed in the infrared and only those to states of *gerade* symmetry will be seen in the Raman; that is no absorptions corresponding to fundamental frequencies of vibration will be common to Raman and infrared spectra. The absence of such coincidences is good (but, of course, negative) evidence for the presence of a centre of symmetry in a molecule: by the same token the presence of absorption bands of the same frequency in both spectra argues against a centre of symmetry.

Table 39

Molecule	Frequencies (mm^{-1})				Conclusion
OCO $\begin{cases} \text{infrared active} \\ \text{Raman active} \end{cases}$	66·7 $-$		$-$ 132	235 $-$	No coincidences, centre of symmetry
NO$_3^-$ $\begin{cases} \text{infrared active} \\ \text{Raman active} \end{cases}$	68 69	83 $-$	$-$ 104·9	135 135·5	Coincidences, no centre of symmetry

(See also problem 7.1.)

7.2.1 Water, H_2O (C_{2v})

The three normal vibrations $Q_1(v_1)$, $Q_2(v_2)$ and $Q_3(v_3)$ have representations A_1, A_1 and B_1. The complete vibrational representation is

$$\Gamma(\Psi) = \Gamma(\psi_1)\Gamma(\psi_2)\Gamma(\psi_3)$$

$$= (\Gamma Q_1)^{n_1}(\Gamma Q_2)^{n_2}(\Gamma Q_3)^{n_3}.$$

If Ψ_I is the ground state then $n_1 = n_2 = n_3 = 0$ and so $\Gamma(\Psi_I) = A_1$. When a single quantum v_1 is absorbed $n_1 = 1$, $n_2 = 0$, $n_3 = 0$; or a single quantum v_2, $n_1 = 0$, $n_2 = 1$, $n_3 = 0$, $\Gamma(\Psi_{II}) = A_1$ since $\Gamma(Q_1) = \Gamma(Q_2) = A_1$. A_1 is also the representation for states that would result from multiple absorption of either or both quanta. The direct product $\Gamma(\Psi_I)\Gamma(\Psi_{II}) = A_1$, but the representations for $(\mathscr{P}, \text{i.r.})$ are $\Gamma_x = B_1$, $\Gamma_y = B_2$ and $\Gamma_z = A_1$ so that the right-hand side of equation 7.2 is only rendered wholly symmetric by the z-component of the electric vector of the infrared radiation. Indeed, if the light were polarized in the xy plane neither of the A_1 frequencies would be observed. Light of frequency v_3 can excite the B_1

vibration. If one quantum is absorbed $n_1 = 0, n_2 = 0, n_3 = 1$ and $\Gamma(\Psi_{II}) = B_1$. If the vibration is doubly excited $n_1 = 0, n_2 = 0, n_3 = 2$ and $\Gamma(\Psi_{II}) = B_1^2 = A_1$; for odd values of n_3 the representation is B_1, for even values A_1. Odd values of n_3 are therefore excited by the x-vector but even values by the z-vector. Nonpolarized infrared light will therefore excite all the fundamental vibrations of water and also all possible overtone and combination bands. Light polarized parallel to the z-axis would excite Q_1 and Q_2 and all their overtones but only the even overtones of Q_3, x-polarized light would interact only with the fundamental and odd overtones of Q_3 and y-polarized light would not be absorbed by water at all.

The representations for $(\mathscr{P}, R)$ are $\Gamma(x^2) = \Gamma(y^2) = \Gamma(z^2) = A_1$, $\Gamma(xy) = A_2$, $\Gamma(xz) = B_1, \Gamma(yz) = B_2$ so that as in the infrared all fundamental frequencies are active as are all possible overtones and combinations.

7.2.2 Cis- *and* trans-*dichloroethylene*, $C_2H_2Cl_2$ (C_{2v} and C_{2h})

Both of these molecules have six atoms and therefore eighteen possible atomic motions. The representations of the twelve fundamental vibrations may be found in the usual way,

trans	C_{2h}	E	C_2	i	σ_h
	$\Gamma(3N)$	18	0	0	6
	Γ_{trans}	3	-1	-3	$+1$ $(x, B_u; y, B_u; z, A_u)$
	Γ_{rot}	3	-1	$+3$	-1 $(R_x, B_g; R_y, B_g; R_z, A_g)$
By difference:	Γ_{vib}	12	$+2$	0	$+6$

Thus $\Gamma_{\text{vib}} = 5A_g + B_g + 2A_u + 4B_u$.

cis	C_{2v}	E	C_2	$\sigma_v(x)$	$\sigma_v(y)$
	$\Gamma(3N)$	18	0	0	6
	Γ_{trans}	3	-1	$+1$	$+1$ $(x, B_1; y, B_2; z, A_1)$
	Γ_{rot}	3	-1	-1	-1 $(R_x, B_2; R_y, B_1; R_z, A_2)$
	Γ_{vib}	12	$+2$	0	$+6$

Thus $\Gamma_{\text{vib}} = 5A_1 + B_2 + 2A_2 + 4B_1$.

In the *trans* molecule only six of the possible fundamental vibrations are infrared active, A_u and B_u, since $(\mathscr{P}, \text{i.r.})$ has representations A_u and B_u. But ten of the twelve vibrational modes $(A_1, B_1$ and $B_2)$ should be infrared active in *cis*-dichloroethylene, $\Gamma(\mathscr{P}, \text{i.r.}) = A_1 + B_1 + B_2$.

Frequencies corresponding to the absorption of even numbers of quanta by the *trans* molecule will not be infrared active since $A_u^{2n} = B_u^{2n} = A_g$; only odd overtones will be active. Combination bands will be of special interest since some of the frequencies of this type result from excitations involving modes whose fundamentals are not infrared active. Thus the simple excitation of any A_g mode in combination with the simultaneous excitation of an A_u mode will have representation $A_g A_u = A_u$ (i.r. active, z-polarized) or an A_g mode with a B_u mode give $A_g B_u = B_u$ (again i.r. active, x-, y-polarized). Similarly $B_g A_u = B_u, B_g B_u = A_u$. A study of combination bands could thus yield information about the frequencies of infrared-inactive vibrations.

All overtones of A_1, B_1 and B_2 vibrations of *cis*-dichloroethylene will be infrared active since A_1 as well as B_1 and B_2 is a component of $\Gamma(\mathscr{P}, \text{i.r.})$. This will also permit the observation of the even overtones of the A_2 vibrations, whose fundamentals (and odd overtones) are infrared forbidden. Combination bands (e.g. B_1 and A_2) can also involve these vibrations.

The Raman spectrum of the *trans* isomer will show none of the same fundamental frequencies as in the infrared spectrum since the molecule has a centre of symmetry. The components of $(\mathscr{P}, R)$ have representations, x^2, y^2, z^2, $xy - A_g$; xz, $yz - B_g$. Thus only A_g and B_g vibrations will be Raman active. Their overtones will also be Raman active as will be the even overtones of the A_u and B_u vibrations. Combination of two Raman-inactive vibrations could yield a Raman-active state (e.g. $A_u B_u = B_g$), but other combinations involving one Raman-active fundamental will not (e.g. $A_g B_u = B_u$). The observation of these bands would be a matter of some difficulty because the intensity, even of Raman fundamentals, is rather weak.

The Raman spectra of the *cis* isomer will be much more complex since all vibrations will be active (x^2, y^2, $z - A_1$; $xy - A_2$; $xz - B_1$; $yz - B_2$). Thus all possible overtones and combinations will also be Raman active.

7.2.3   *Benzene*, C_6H_6 ($\boldsymbol{D_{6h}}$)
The representations for the thirty fundamental vibrations may be determined as follows:

	E	C_2	C_3	C_6	C_2'	C_2''	i	σ_h	S_6	S_3	σ_v	σ_v'
$\Gamma(3N)$	36	0	0	0	−4	0	0	12	0	0	0	0
Γ_{trans}	3	−1	0	+2	−1	−1	−3	+1	0	−2	+1	+1
Γ_{rot}	3	−1	0	+2	−1	−1	+3	−1	0	+2	−1	−1

Hence

Γ_{vib}	30	+2	0	−4	−2	+2	0	+12	0	0	0	+4

Thus $\Gamma_{\text{vib}} = 2A_{1g} + A_{2g} + A_{2u} + 2B_{1u} + 2B_{2g} + 2B_{2u} + E_{1g} + 3E_{1u} + 4E_{2g} + 2E_{2u}$,

also $\Gamma(\mathscr{P}, \text{i.r.}) = A_{2u}(z) + E_{1u}(x, y)$.

$$\Gamma(\mathscr{P}, R) = 2A_{1g}\{(z)^2 \text{ and } (x^2 + y^2)\} + E_{1g}(xz, yz) + E_{2g}(x^2 - y^2, xy).$$

There are no coincidences between Raman and infrared frequencies indicating that benzene has a centre of symmetry. The predictions as to the number of permitted vibrations (4, i.r. − A_{2u} and $3E_{1u}$ and 7, R − $2A_{1g} + E_{1g} + 4E_{2g}$) has been completely confirmed by Ingold *et al.* ('Structure of benezene. Part VIII', *Journal of the Chemical Society*, 1936, paper 218, pp. 971–87; 'Structure of benzene. Parts XI and XVI–XXI', *Journal of the Chemical Society*, 1946, pp. 222–35 and 252–333). This work may be regarded both as a proof of the planar hexagonal structure of benzene (i.e. $\boldsymbol{D_{6h}}$) and also as a demonstration of the power of group theory.

189   Selection Rules for Infrared and Raman Spectra

7.3 Characteristic bond frequencies

In the examples so far considered attention has been focused on the vibration of the molecule as a whole but as can be seen from the first few figures in this chapter vibrations are very often almost wholly associated with one particular bond or a pair of bonds or the angle between a pair of bonds. In many molecules this is because of a disparity in mass between the oscillating nuclei. Thus in a C–H bond the hydrogen will do most of the moving. Indeed it is possible to ascribe particular frequencies to the stretching and bending of bonds which vary little from molecule to molecule. A knowledge of such characteristic bond frequencies is an invaluable aid in determining molecular structures. By way of example the various vibrations associated with the $-CH_2-$ group will be considered; also the $C \equiv O$ frequencies in octahedral complexes.

7.3.1 Methylene, $>CH_2 (C_{2v})$

The local symmetry of this group is C_{2v} but unlike the water molecule it can neither rotate nor translate freely. To a first approximation the carbon atom may be regarded as fixed. The six possible motions of the H atoms must therefore all be considered as contributing to vibrations of the C–H bonds.

	E	C	$\sigma_v(yz)$	$\sigma_v(xz)$
$\Gamma(3N)(2Hs)$	6	0	0	$+2$

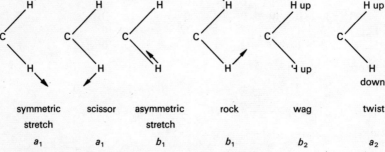

Figure 83   The vibrations of the methylene group

These vibrations have representations $2A_1 + A_2 + 2B_1 + B_2$ (Figure 83). The fundamentals of all these are infrared active save for A_2.

7.3.2 Carbonyl complexes

Six carbonyl groups around a transition metal give rise to an octahedral complex (e.g. $Mo(CO)_6$).

The characteristic stretching frequency of the carbonyl group in such complexes does not vary much from compound to compound. However, when there are six carbonyl groups the six stretching frequencies should be considered together under the symmetry operations of the molecule: they will be A_{1g}, E_g and T_{1u}. Since the fundamental vibration is the same for all representations, they should be energetically degenerate, but interactions through the central atom will separate the frequencies somewhat. In this case (O_h) only one of the three possible representations will be i.r. active (T_{1u}) since ($\mathscr{P}$, i.r.) is T_{1u}. Some other possible carbonyl complexes in which various carbonyls have been replaced by other ligands, their point groups, and the number of infrared absorptions to be expected in the carbonyl region, are shown in Table 40. Figure 84 shows the number of observed absorptions for various complexes. Such information will obviously be of value in elucidating the structure of a complex species.

Table 40

Molecule	Point group	Representations for CO stretches	($\mathscr{P}$, i.r.)	Number of infrared active frequencies
$M(CO)_6$	O_h	$A_{1g}+E_g+T_{1u}$	T_{1u}	1
$MX(CO)_5$	C_{4v}	$2A_1+B_1+E$	A_1, E	3
(structure: D_{4h}, $MX_2(CO)_4$)	D_{4h}	$A_{1g}+B_{1g}+E_u$	A_{2u}, E_u	1
(structure: C_{2v}, $MX_2(CO)_4$)	C_{2v}	$2A_1+B_1+B_2$	A_1, B_1, B_2	4
(structure: C_{3v}, $MX_3(CO)_3$)	C_{3v}	A_1+E	A_1, E	2
(structure: C_{2v}, $MX_3(CO)_3$)	C_{2v}	$2A_1+B_1$	A_1, B_1, B_2	3

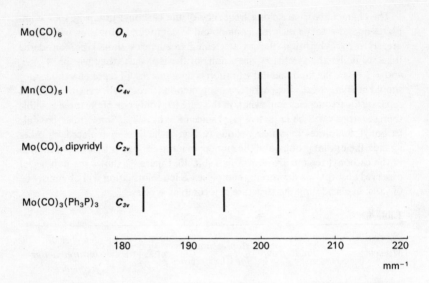

Figure 84   Infrared absorptions of some molybdenum carbonyl complexes

7.4   Crystallography

One of the most obvious manifestations of symmetry in chemistry is in the shape of crystals. Indeed it is to crystallographers that we are indebted for having developed the schemes of classification by symmetry elements described in Chapter 3. Unfortunately, there are now in wide use two systems of nomenclature, both for elements of symmetry and symmetry groups. The system used so far in this book is that developed by Schönflies (S). However the Hermann–Mauguin (H–M) is preferred by crystallographers as being more systematic and more easily extended to describe space groups.

In the description of the Hermann–Mauguin notation that follows, the corresponding Schönflies symbol will be given in brackets. Axes of rotation through $(360/n)°$ are symbolized by the number n, thus $1(C_1)$, $2(C_2)$, $3(C_3)$, $4(C_4)$, $6(C_6)$.

The Schönflies symmetry operation of improper rotation (S_n) is not used in the Hermann–Mauguin notation. Instead a new element of symmetry is defined, the inversion axis. The corresponding symmetry operation is a rotation through $(360/n)°$ followed by inversion, and is given the symbol $\bar{n}$.

Thus $\bar{1}\,(i \equiv S_2)$ $\bar{2}\,(\sigma_h \equiv S_1)$ $\bar{3}\,(S_6^5)$ $\bar{4}\,(S_4^3)$ $\bar{6}\,(S_3^5)$.

Mirror planes are indicated by m, but if the plane is perpendicular to the principal axis (σ_h), the symbol n/m is used. The various point groups may be defined by listing their symmetry elements. The symbol for the principal axis of rotation is placed first, followed by such other symmetry elements as are necessary for a unique (if not complete) description. As will be shown later only thirty-two point groups are of crystallographic importance and consequently considerable abbreviation of the group symbol is possible without causing confusion. Thus D_{2h} has three twofold axes at right angles to each other and perpendicular to each axis, a mirror plane. The full symbol would be 2/m2/m2/m but this may be written mmm since three mutually perpendicular planes of reflection will generate the twofold axes. Table 41 relates the symbols for crystallographic point groups in the two notations.

Table 41

	H–M	S		H–M	S
Triclinic	1	C_1	Tetragonal	4	C_4
	$\bar{1}$	$C_i\,(S_2)$		4mm	C_{4v}
				4/m	C_{4h}
Monoclinic	2	C_2		422	D_4
	2/m	C_{2h}		4/mmm	D_{4h}
	m	C_{1h}		$\bar{4}$2m	D_{2d}
				$\bar{4}$	S_4
Orthorhombic	2mm	C_{2v}			
	222	D_2	Hexagonal	6	C_6
	mmm	D_{2h}		6mm	C_{6v}
				$\bar{6}$	C_{3h}
Trigonal	3	C_3		6/m	C_{6h}
	3m	C_{3v}		622	D_6
	32	D_3		$\bar{6}$m2	D_{3h}
	$\bar{3}$m	D_{3d}		6/mmm	D_{6h}
	$\bar{3}$	$C_{3i},\,S_6$			
			Cubic	$\bar{4}$3m	T_d
				23	T
				m3	T_h
				434	O
				m3m	O_h

Figure 85 shows diagrams of some crystals with an indication of the symmetry groups or *crystal classes* to which they belong.

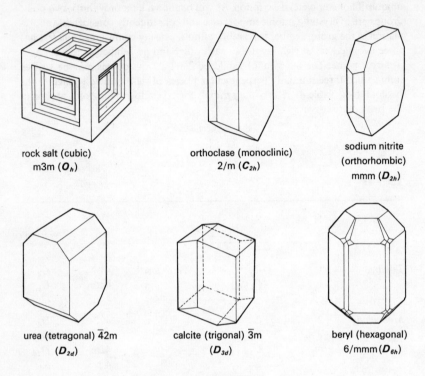

rock salt (cubic)
m3m (O_h)

orthoclase (monoclinic)
2/m (C_{2h})

sodium nitrite
(orthorhombic)
mmm (D_{2h})

urea (tetragonal) $\bar{4}$2m
(D_{2d})

calcite (trigonal) $\bar{3}$m
(D_{3d})

beryl (hexagonal)
6/mmm (D_{6h})

Figure 85   Crystal structures of some well-known compounds

7.5   The structure of crystals

The symmetrical appearance of crystals reflects their underlying structure – a regular arrangement in space of molecules or ions.

It will be possible to find some basic pattern unit which by continuous repetition in three dimensions will reproduce the whole crystal structure, leaving no gaps. Such a pattern unit may by itself constitute a *unit cell*, or for convenience two or more pattern units may be considered together as the unit cell. The choice of unit cell is rather arbitrary and is usually determined by the requirements and conventions of X-ray crystallography. As with the basic pattern unit it is required of the unit cell that it should be possible by simple repetition of this unit, with the same orientation in space to build up the whole crystal. This places certain

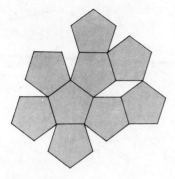

Figure 86   The inability of pentagons to cover a plane completely

stringent requirements upon the unit cell's symmetry properties and also upon the symmetry properties of the resulting crystal. For instance, the cell *cannot* possess a fivefold axis of symmetry because it is not possible to pack regular pentagons together without leaving gaps (Figure 86). The permitted elements of rotational symmetry may be found as follows. Let each dot in Figure 87 represent a unit cell. The figure thus represents part (a two-dimensional layer) of the array of unit cells that makes up a crystal. Consider rotations perpendicular to the layer of dots. If any rotation is to be an element of symmetry then it must restore

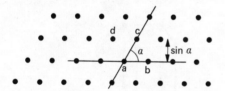

Figure 87   Two-dimensional crystal lattice

Figure 87 to its original appearance. Suppose the axis passes through point a, rotation anticlockwise through the angle α will take b to c, c to d, etc., and also all other points in row ab into row ac, etc. For this to be possible the distance between adjacent rows must be sin α (given that the distance between any two points in any row is unity). If the operation is repeated point b, now at site c, is carried to site d. Since the rows are equally spaced sin 2α, the length of the perpendicular from d to row ab, must be a multiple of sin α,

i.e. $\sin 2\alpha = n \sin \alpha = 2 \sin \alpha \cos \alpha$

$\frac{1}{2}n = \cos \alpha$

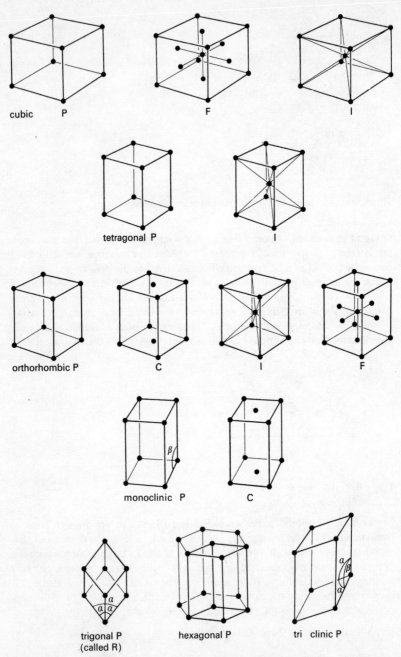

cubic   P

F

I

tetragonal P

I

orthorhombic P

C

I

F

monoclinic P

C

trigonal P
(called R)

hexagonal P

tri clinic P

Figure 88   Unit cells consistent with the Bravais lattices

But $-1 \leqslant \cos\alpha \leqslant +1$ so that n can only have values $\pm 2, \pm 1, 0$. The corresponding values of α are $0°, 180°, 60°, 120°, 90°$. Thus the only rotational elements of symmetry that are permitted in crystals, if the unit cells are to pack properly, are $1(C_1 \equiv E), 2(C_2), 3(C_3), 4(C_4)$ and $6(C_6)$. Reflections in mirror planes and inversion through a centre of symmetry are also possible operations; admitting them results in the thirty-two different symmetry point groups in Table 40. These groups have been arranged in the conventional crystal systems.

The ways in which the unit cells of different crystals may be arranged in space depend on the actual shape of the cell. Such spatial arrangements are *lattices*. For simplicity lattices may be thought of as three-dimensional networks of *lattice points*. The environment of each lattice point should be identical. It is found that just fourteen such lattices exist, the Bravais lattices. Unit cells consistent with these lattices are shown in Figure 88. In the first cell, cubic P, lattice points are at the eight corners of the unit cell. Clearly repetition of such a cube in three dimensions would generate a lattice in which all the lattice points would be identical (as required). Also since each lattice point is shared by eight unit cells, each unit cell is equivalent to one lattice point. In the second example, cubic F, lattice points are to be found at the centre of each face of the unit cell (*face-centred* cubic), again the repetition of such a unit cell will generate a lattice in which all the points have the same surroundings. The six lattice points in the cube faces are shared between two unit cells so that the cubic F unit cell represents four lattice points altogether (quadruply primitive); similar arguments will show that the cubic I-cell is doubly primitive. In the F and I cases it would be possible to construct a primitive cell representing just one lattice point but the visual relationship to the P-cubic lattice would be lost.

Tetragonal crystals have the common feature of a fourfold axis and this is retained in the two possible tetragonal lattices, P and I. The edges of ortho-rhombic unit cells and crystals meet at right angles but only twofold axes of

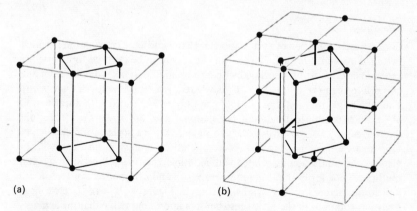

(a) (b)

Figure 89 'New' lattices: (a) the 'tetragonal C' is the same as the tetragonal P; (b) the 'tetragonal F' is the same as the tetragonal I

rotation are present. The four possible lattices are P, C (a pair of opposite faces of the unit cell have lattice points at their centres), I and F. It is now worth reflecting on why there are no tetragonal cells that correspond to the orthorhombic C and F. Suppose the square faces of the tetragonal unit cells were to contain lattice points so as to generate a C-lattice, then as can be seen from Figure 89 the 'new' lattice is still in fact only a tetragonal P-lattice (the size of the square faces is smaller and the unit cell has been rotated through 45° but it is not a new lattice). Similarly the tetragonal 'F-lattice' is the same as the tetragonal I-lattice. The other possible lattices compatible with unit cells from other crystal classes are also shown in Figure 88.

7.5.1 Translational symmetry operations

It is important to realize that the unit cells in the diagrams above may well represent a large number of atoms, ions or molecules (or parts of molecules – the unit cell is merely the repeating pattern unit and is no respecter of chemical bonds). The way in which the constituent parts are arranged in the unit cell determines the over-all symmetry of the cell. One might well seek to find the symmetry of a unit cell by determining to what point group it belongs but this treats the cell as an isolated unit and ignores a vital point of crystal structure: that the unit cell is not alone but surrounded on all sides by identical neighbours. Granted an 'infinite' (for the purpose of this argument) array of repeating units, it is then possible to admit of new symmetry operations which involve movement in space, from one cell to another. For example, a translation of the whole structure by the length of the edge of one unit cell and parallel to that edge would have the same appearance as the original crystal. This simple operation may be thought of as the identity operation of a point group coupled with a translation (over a distance t) and may be symbolized Et. The translation operation may also be coupled with other symmetry operations, rotations and reflections.

7.5.2 Screw axis

Consider a combination of a translation and a twofold axis. The rotation repeated twice is a rotation through 360°, so for the crystal to be restored to its original appearance C_2^2 must be associated with a translation that takes a point from one unit cell to the next, that is $Ea = C_2^2 a$ (where a is a typical unit cell dimension). Thus the single rotation through 180° can only be associated with a translation of $\frac{1}{2}a$. Such combined rotational–translational axes are screw axes. Threefold screw axes may also be envisaged: here, of course, the associated translation per 120° turn can only be $\frac{1}{3}a$ or $\frac{2}{3}a$. These operations are shown in Figure 90. In general, the translation associated with an n-fold rotation will be some integral multiple of a/n, for example $x(a/n)$ where $x < n$ and a, as above, is a typical cell dimension. Screw axes are symbolized n_x. In Figure 90, 3_1 and 3_2 have been drawn as rotations in the same sense but it is interesting to see that the arrangement of points in space for 3_2 clockwise is the same as for 3_1 anticlockwise. Similarly, 4_3, 6_5 and 6_4 are mirror images of 4_1, 6_1 and 6_2 respectively.

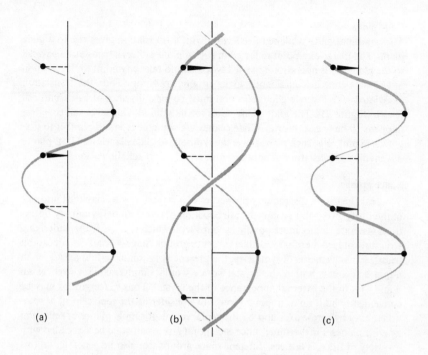

Figure 90   Screw axes: (a) 3_1 clockwise; (b) 3_2 clockwise; (c) 3_1 anticlockwise. Note that the arrangement of points is the same in (b) and (c)

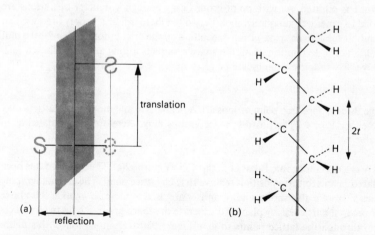

Figure 91   (a) Diagram illustrating glide reflection. (b) Glide reflection in a saturated hydrocarbon chain.

7.5.3 Glide plane

The combination of a plane of reflection with a translation gives rise to a glide plane. The image of a point reflected in such a plane suffers a translation parallel to the plane. This operation repeated twice should take any point to the position in the next unit cell analogous to its original position. Thus the translation associated with a single glide reflection must be for a distance of half a unit cell length (Figure 91). The glide plane takes for its symbol whichever cell edge this happens to be (e.g. a, b or c). Glide planes are sometimes found parallel to diagonals in unit cells, such planes have the symbol n (d glide planes are like n-planes but involve a translation of only one quarter of a unit cell dimension).

7.6 Space groups

Whereas the external symmetry properties of a crystal may be classified into one of the thirty-two point groups the introduction of translational symmetry operations generates many more possible groups with which to classify the underlying structure. Indeed there are 230 different space groups. Since crystals cannot exhibit translational elements of symmetry, the presence of a simple n-fold axis *or* of an n-fold screw axis within the crystal lattice would simply manifest itself as an n-fold axis in the external appearance of the crystal. Thus space groups may be conveniently built up from point groups by considering the replacement of rotational axes by screw axes and by the replacement of simple planes of reflection by glide planes. Furthermore, since each point group can often be associated with a variety of Bravais lattices, different space groups can also be generated in this way. In naming space groups, the form of the lattice is given first by a capital letter, e.g. P, C, I, F, etc., then follows the principal axes, whether it be simple, inversion or screw. If a plane of reflection or a glide plane is perpendicular to this axis, the symbol is /m, /a, or /b, etc., whereas if the plane contains the axis the stroke is omitted. As with point groups only enough symmetry operations are listed to generate a unique symbol. Very often there will be a variety of equivalent symbols for a given space group depending on the orientation chosen for the unit cell. Since a description of all the space groups is a long and complicated task only a few simple examples will be given here.

7.6.1 Monoclinic space groups

The three monoclinic point groups are 2, m and 2/m and monoclinic lattices may be either P or C. The possible space groups may, therefore, be tabulated as in Table 42.

A typical monoclinic P-lattice is shown in Figure 92(a). The symbol S has been placed in an arbitrary position relative to each lattice point. The crystallographic axes x, y, z are also shown and + and − are used to indicate whether S is above or below a particular xy plane. To generate the space group P2, let twofold axes run through the lattice points to give Figure 92(b). Examination of this figure reveals that there will be additional twofold axes in this space group – they are marked with open barbs. If twofold screw axes were to pass through the lattice

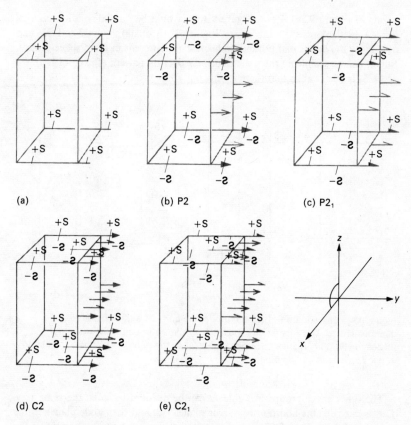

(a)   (b) P2   (c) P2₁

(d) C2   (e) C2₁

Figure 92   (a) Typical monoclinic P-lattice; (b) the addition of twofold symmetry axes yields a P2 space group; (c) the $P2_1$ space group, formed by the addition of twofold screw axes; (d) the C2 and (e) the $C2_1$ space groups, showing the normal and screw axes of twofold symmetry

Table 42

	2		m		2/m	
P	P2	$P2_1$	Pm	Pc	P2/m	$P2_1$/m
					P2/c	$P2_1$/c
C	C2	$C2_1$	Cm	Cc	C2/m	$C2_1$/m
					C2/c	$C2_1$/c

Not all these symbols are unique, $C2 = C2_1$, $C2/m = C2_1/m$ and $C2/c = C2_1/c$.

points of Figure 92(a), P2$_1$ – Figure 92(c) – would be formed; additional axes (screw) are also present in this group. Figures 92(d) and (e) show C2 and C2$_1$ and also show the additional normal and screw axes present in these space groups. But, as can be seen from these diagrams, the over-all pattern of axes for both C2 and C2$_1$ is identical so that there is only one space group.

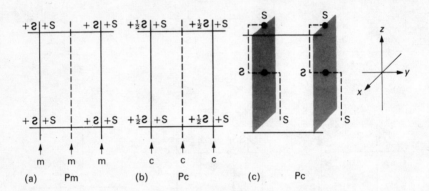

Figure 93   The addition of (a) a mirror plane and (b) a glide plane to a primitive lattice. The illustrations show the *xy* planes in unit cells: the mirror planes m and glide planes c are perpendicular to the paper. (c) Shows the action in the *yz* plane of a glide plane

The other space groups in Table 42 can be deduced in much the same way. Let us consider the addition of mirror planes (*xz*) and then glide planes to the primitive lattice. Figure 93(a) shows the *xy* plane at the top or bottom of the unit cell. Notice that the array of Ss generated for Pm also requires the presence of another mirror plane parallel to the *xz* faces of the unit cell, halfway across. Figure 93(b) shows the effect of combining a glide plane with a P-lattice. The direction of the translation associated with the glide plane is along the *z*-axis. If the unit cell is orientated so that the c-edge is parallel to this axis the lattice generated is a Pc lattice. The mirror images of the Ss of the P-lattice will now be half way up the edge of the unit cell symbolized by $+\frac{1}{2}2$ in the diagram. Repetition of this operation will of course generate a $+$S at the bottom of the next unit cell. It should be noticed that just as in Pm a new mirror plane was automatically generated in the middle of the unit cell so in the space group Pc a new c glide plane is present in the middle of the cell parallel to the original glide planes.

Careful considerations of this type permit all the other monoclinic space groups to be generated. It will be found that the presence of the *essential* symmetry operations will necessarily generate other operations. Since the array of twofold screw and normal axes for C2 and C2$_1$ is in fact the same, it will be found that the space groups C2/m and C2$_1$/m are also identical as are C2/c and C2$_1$/c.

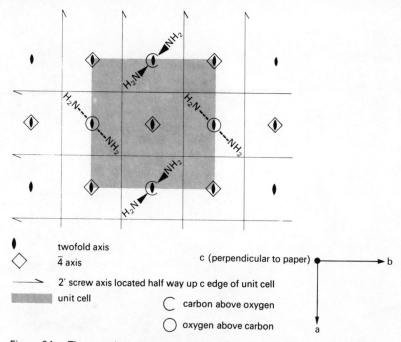

twofold axis

◇ $\bar{4}$ axis   c (perpendicular to paper)

⟶ 2′ screw axis located half way up c edge of unit cell

▨ unit cell

C carbon above oxygen

O oxygen above carbon

b

a

Figure 94   The crystal structure (tetragonal P$\bar{4}2_1$m) of urea. Four molecules are shown. The C=O bond is perpendicular to the plane of the paper. Alternate molecules show carbon and oxygen atoms uppermost

7.6.2   Urea (tetragonal – P$\bar{4}2_1$m)

As an example of a rather more complex space group in another crystal class, consider the crystal structure of urea. Urea, $(NH_2)_2CO$, crystallizes in tetragonal crystals as shown in Figure 85, p. 194. These crystals have symmetry $\bar{4}2m(D_{2d})$. One of the space groups that may be derived from this point group, P$\bar{4}2_1$m, corresponds to the arrangement of urea molecules in the crystal. Figure 94 shows the location of symmetry operations for this space group. The mirror planes of this group intersect at right angles (tetragonal system) and so where they cut must contain twofold axes. The local symmetry about any point on such an axis could be mm2 (C_{2v}). Urea has this symmetry (ignoring the hydrogens) and so urea molecules might well be found where the m-planes in P$\bar{4}2_1$m intersect. Such sites are in fact occupied in this way (Figure 94). Notice also how the urea molecules are inter-related by the screw axes, the fourfold inversion axes, and the glide planes.

Problem 7

7.1   Deduce the number of independent fundamental modes of vibration for CO_2 and NO_3^-. Determine which will be infrared active and which Raman active.

Bibliography

Chapters 1 and 2

General

C. A. COULSON, *Valence*, 2nd Edn, Oxford University Press, 1961.
J. N. MURRELL, S. F. A. KETTLE and J. M. TEDDER, *Valence Theory*, Wiley, 1965.
J. W. LINNETT, *Wave Mechanics and Valency*, Methuen, 1960.

Wave mechanics and quantum mechanics

W. HEITLER, *Elementary Wave Mechanics*, Oxford University Press, 1956.
M. W. HANNA, *Quantum Mechanics in Chemistry*, Benjamin, 1966.
H. F. HAMEKA, *Introduction to Quantum Theory*, Harper International, 1967.
H. EYRING, J. WALTER and G. E. KIMBALL, *Quantum Chemistry*, Wiley, 1944.
L. PAULING and E. B. WILSON, *Introduction to Quantum Mechanics*, McGraw-Hill, 1935.
P. T. MATTHEWS, *Introduction to Quantum Mechanics*, McGraw-Hill, 1963.

Electron correlation

P. G. DICKENS and J. W. LINNETT, 'Electron correlation and chemical consequences', *Quarterly Reviews*, vol. 11 (1957), no. 4, pp. 291–312.

Molecular orbital theory

J. D. ROBERTS, *Notes on Molecular Orbital Calculations*, Benjamin, 1962.
C. J. BALLHAUSEN and H. B. GRAY, *Molecular Orbital Theory*, Benjamin, 1965.
A. STEITWIESER, Jnr, *Molecular Orbital Theory for Organic Chemists*, Wiley, 1961.
L. SALEM, *The Molecular Orbital Theory of Conjugated Systems*, Benjamin, 1966.

Chapters 3 and 4

R. S. MULLIKEN, 'Electronic Structures of Polyatomic Molecules and Valence, IV. Electronic States, Quantum Theory of the Double Bond', *Physical Review*, vol. 43 (1933), pp. 279–302.

G. M. BARROW, *Introduction to Molecular Spectroscopy*, Chapter 8, McGraw-Hill, 1962.

J. E. ROSENTHAL and G. M. MURPHY, 'Group Theory and the Vibrations of Polyatomic Molecules', *Reviews of modern Physics*, vol. 8 (1936), pp. 317–46.

F. A. COTTON, *Chemical Applications of Group Theory*, Interscience, 1964.

H. A. JAFFÉ and M. ORCHIN, *Symmetry in Chemistry*, Wiley, 1965.

D. S. SCHONLAND, *Molecular Symmetry*, Van Nostrand, 1965.

M. TINKHAM, *Group Theory and Quantum Mechanics*, McGraw-Hill, 1964.

V. HEINE, *Group Theory in Quantum Mechanics*, Pergamon, 1964.

R. MCWEENEY, *Symmetry: an Introduction to Group Theory and its Applications*, Pergamon, 1964.

E. P. WIGNER, *Group Theory and its Applications to the Quantum Mechanics of Atomic Spectra*, Academic Press, 1959.

H. C. LONGUET-HIGGINS, 'The symmetry groups of non-rigid molecules', *Journal of molecular Physics*, vol. 6 (1963), pp. 445–60.

Chapter 5

Hybridization

G. E. KIMBALL, 'Directed Valence', *Journal of Chemical Physics*, vol. 8 (1940), pp. 188–98.

Molecular orbital and valence bond theories

J. H. VAN VLECK and A. SHERMAN, *Reviews of modern Physics*, vol. 7 (1935), pp. 167–228.

Examples

H. C. LONGUET-HIGGINS and M. DE V. ROBERTS, 'The electronic structure of an icosahedron of boron atoms', *Proceedings of the Royal Society*, vol. A230 (1955), 110.

G. B. GILL, 'The application of the Woodward-Hoffmann Symmetry Rules to Concerted Organic Reactions', *Quarterly Reviews*, vol. 22 (1968), no. 3, 338-89.

Chapter 6

Electronic states of atoms

G. HERZBERG, *Atomic Spectra and Atomic Structure*, Dover, 1944.

Crystal field and ligand field theories

J. G. GRIFFITH and L. E. ORGEL, 'Ligand-field theory', *Quarterly Reviews*, vol. 11 (1957), no. 4, pp. 381–93.
L. E. ORGEL, *An Introduction to Transition Metal Chemistry: Ligand Field Theory*, Methuen, 1960.
C. J. BALLHAUSEN, *Introduction to Ligand Field Theory*, McGraw-Hill, 1962.

Chapter 7

Infrared and Raman spectra

G. HERZBERG, *Infra-red and Raman Spectra of Polyatomic Molecules*, Van Nostrand, 1946.

Molecular vibrations

J. E. ROSENTHAL and G. M. MURPHY, 'Group Theory and the Vibrations of Polyatomic Molecules', *Reviews of modern Physics*, vol. 8 (1936), pp. 317–46.
E. B. WILSON, J. C. DECIUS and P. C. CROSS, *Molecular Vibrations: the Theory of Infra-red and Raman Vibrational Spectra*, McGraw-Hill, 1955.

Crystallography

F. C. PHILLIPS, *An Introduction to Crystallography*, 3rd Edn, Longmans, 1963.
C. W. BUNN, *Chemical Crystallography*, 2nd Edn, Oxford University Press, 1961.

Space groups

M. J. BUERGER, *Elementary Crystallography*, Wiley, 1956.
G. F. KOSTER, 'Space Groups and their Representations', *Solid State Physics*, vol. 5 (1957), pp. 173–256, Academic Press.

Appendix A
Matrix Multiplication

Let there be two vectors, x_1 and x_2. Suppose there is an operator A whose action upon these vectors can be described by the following equations:

$$A(x_1) = y_1 = a_{11} x_1 + a_{12} x_2,$$

$$A(x_2) = y_2 = a_{21} x_1 + a_{22} x_2.$$

Operator A can be represented by the matrix $\begin{bmatrix} a_{11} & a_{12} \\ a_{21} & a_{22} \end{bmatrix}$.

Now consider an operator B to act upon the new vectors y_1 and y_2,

$$B(y_1) = z_1 = b_{11} y_1 + b_{12} y_2,$$

$$B(y_2) = z_2 = b_{21} y_1 + b_{22} y_2.$$

Thus B can be represented by $\begin{bmatrix} b_{11} & b_{12} \\ b_{21} & b_{22} \end{bmatrix}$.

The over-all operation of A and then B can be written:

$$B.A(x_1) = b_{11}(a_{11} x_1 + a_{12} x_2) + b_{12}(a_{21} x_1 + a_{22} x_2),$$

$$B.A(x_2) = b_{21}(a_{11} x_1 + a_{12} x_2) + b_{22}(a_{21} x_1 + a_{22} x_2).$$

Collecting coefficients of x_1 and x_2, the matrix corresponding to the product of operation B and A is

$$B.A\begin{bmatrix} x_1 \\ x_2 \end{bmatrix} = \begin{bmatrix} b_{11} a_{11} + b_{12} a_{21} & b_{11} a_{12} + b_{12} a_{22} \\ b_{21} a_{11} + b_{22} a_{21} & b_{21} a_{12} + b_{22} a_{22} \end{bmatrix} \begin{bmatrix} x_1 \\ x_2 \end{bmatrix} = \begin{bmatrix} c_{11} & c_{12} \\ c_{21} & c_{22} \end{bmatrix} \begin{bmatrix} x_1 \\ x_2 \end{bmatrix}.$$

The coefficients c form the matrix that represents BA.

This example may be generalized so that the coefficients c may be found in terms of as and bs,

$$c_{ij} = \sum_{l=1}^{n} b_{il} a_{lj},$$

where n is the number of rows or columns in the matrix. The first number in the subscript indicates a row. The second number in the subscript indicates a column.

Thus $c_{11} = \sum\limits_{}^{2} b_{1l} a_{l1}$. This may be demonstrated on the matrices for B and A:

$$\begin{array}{c} \text{column 1} \\ l = 1 \quad (a_{l1}) \end{array}$$

$$\text{row 1 } (b_{1l}) \begin{bmatrix} b_{11} & b_{12} \\ b_{21} & b_{22} \end{bmatrix} \begin{bmatrix} a_{11} & a_{12} \\ a_{21} & a_{22} \end{bmatrix}.$$
$$l = 2$$

Similarly for $c_{12} = \sum\limits_{}^{2} b_{1l} a_{l2}$:

$$\begin{array}{c} \text{column 2} \\ l = 1 \quad (a_{l2}) \end{array}$$

$$\text{row 1} (b_{1l}) \begin{bmatrix} b_{11} & b_{12} \\ b_{21} & b_{22} \end{bmatrix} \begin{bmatrix} a_{11} & a_{12} \\ a_{21} & a_{22} \end{bmatrix}.$$
$$l = 2$$

For $c_{21} = \sum\limits_{}^{2} b_{2l} a_{l1}$:

$$\begin{array}{c} \text{column 1} \\ l = 1 \quad (a_{l1}) \end{array}$$

$$\text{row 2 } (b_{2l}) \begin{bmatrix} b_{11} & b_{12} \\ b_{21} & b_{22} \end{bmatrix} \begin{bmatrix} a_{11} & a_{12} \\ a_{21} & a_{22} \end{bmatrix},$$
$$l = 2$$

etc.

Appendix B
Character Tables for Point Groups

C_1	E
A	1

$C_{1h} \equiv C_s$	E	σ_h		
A'	1	1	x, y, xy, x^2, y^2, z^2	R_z
A''	1	-1	z, xz, yz	R_x, R_y

$S_2 \equiv C_i$	E	$C_2\,\sigma_h = S_2 = i$		
A_g	1	1	$xy, xz, yz, x^2, y^2, z^2$	R_x, R_y, R_z
A_u	1	-1	x, y, z	

Twofold axis

C_2	E	C_2		
A	1	1	z, xy, x^2, y^2, z^2	R_z
B	1	-1	x, y, xz, yz	R_x, R_y

C_{2v}	E	C_2	$\sigma_v(xz)$	$\sigma_v'(yz)$		
A_1	1	1	1	1	z, x^2, y^2, z^2	
A_2	1	1	-1	-1	xy	R_z
B_1	1	-1	1	-1	x, xz	R_y
B_2	1	-1	-1	1	y, yz	R_x

C_{2h}	E	C_2	σ_h	i		
A_g	1	1	1	1	xy, x^2, y^2, z^2	R_z
A_u	1	1	-1	-1	z	
B_g	1	-1	-1	1	xz, yz	R_x, R_y
B_u	1	-1	1	-1	x, y	

D_2	E	$C_2(z)$	$C_2(y)$	$C_2(x)$		
A_1	1	1	1	1	x^2, y^2, z^2	
B_1	1	1	-1	-1	z, xy	R_z
B_2	1	-1	1	-1	y, xz	R_y
B_3	1	-1	-1	1	x, yz	R_x

D_{2h}	E	$C_2(z)$	$C_2(y)$	$C_2(x)$	i	$\sigma_h(xy)$	$\sigma_h(xz)$	$\sigma_h(yz)$		
A_{1g}	1	1	1	1	1	1	1	1	x^2, y^2, z^2	
A_{1u}	1	1	1	1	-1	-1	-1	-1		
B_{1g}	1	1	-1	-1	1	1	-1	-1	xy	R_z
B_{1u}	1	1	-1	-1	-1	-1	1	1	z	
B_{2g}	1	-1	1	-1	1	-1	1	-1	xz	R_y
B_{2u}	1	-1	1	-1	-1	1	-1	1	y	
B_{3g}	1	-1	-1	1	1	-1	-1	1	yz	R_x
B_{3u}	1	-1	-1	1	-1	1	1	-1	x	

S_4	E	C_2	S_4	S_4^3		
A	1	1	1	1	z^2, x^2+y^2	R_z
B	1	1	-1	-1	z, x^2-y^2, xy	
E	$\left\{\begin{matrix} 1 \\ 1 \end{matrix}\right.$	$\begin{matrix} -1 \\ -1 \end{matrix}$	$\begin{matrix} i \\ -i \end{matrix}$	$\left.\begin{matrix} -i \\ i \end{matrix}\right\}$	$(x, y), (xz, yz)$	(R_x, R_y)

$D_{2d}(S_{4v})$	E	C_2	$2S_4$	$2C_2'$	$2\sigma_d$		
A_1	1	1	1	1	1	z^2, x^2+y^2	
A_2	1	1	1	-1	-1		R_z
B_1	1	1	-1	1	-1	x^2-y^2	
B_2	1	1	-1	-1	1	z, xy	
E	2	-2	0	0	0	$(x, y), (xz, yz)$	(R_x, R_y)

Threefold axis

C_3	E	C_3	C_3^2		
A	1	1	1	z, z^2, x^2+y^2	R_z
E	$\begin{cases} 1 \\ 1 \end{cases}$	$\begin{matrix} \omega \\ \omega^2 \end{matrix}$	$\left.\begin{matrix} \omega^2 \\ \omega \end{matrix}\right\}$	$\begin{matrix} (x, y), (xz, yz) \\ (x^2-y^2, xy) \end{matrix}$	(R_x, R_y)

$$\omega = \exp\frac{2\pi i}{3}$$

C_{3v}	E	$2C_3$	$3\sigma_v$		
A_1	1	1	1	z, z^2, x^2+y^2	
A_2	1	1	-1		R_z
E	2	-1	0	$\left.\begin{matrix} (x, y), (xz, yz) \\ (x^2-y^2, xy) \end{matrix}\right\}$	(R_x, R_y)

C_{3h}	E	C_3	C_3^2	σ_h	S_3	$C_3^2\,\sigma_h \equiv S_3^5$		
A'	1	1	1	1	1	1	z^2, x^2+y^2	R_z
A''	1	1	1	-1	-1	-1	z	
E'	$\begin{cases} 1 \\ 1 \end{cases}$	$\begin{matrix} \omega \\ \omega^2 \end{matrix}$	$\begin{matrix} \omega^2 \\ \omega \end{matrix}$	$\begin{matrix} 1 \\ 1 \end{matrix}$	$\begin{matrix} \omega \\ \omega^2 \end{matrix}$	$\left.\begin{matrix} \omega^2 \\ \omega \end{matrix}\right\}$	$\begin{matrix} (x, y) \\ (x^2-y^2, xy) \end{matrix}$	
E''	$\begin{cases} 1 \\ 1 \end{cases}$	$\begin{matrix} \omega \\ \omega^2 \end{matrix}$	$\begin{matrix} \omega^2 \\ \omega \end{matrix}$	$\begin{matrix} -1 \\ -1 \end{matrix}$	$\begin{matrix} -\omega \\ -\omega^2 \end{matrix}$	$\left.\begin{matrix} -\omega^2 \\ -\omega \end{matrix}\right\}$	(xz, yz)	R_x, R_y

$$\omega = \exp\frac{2\pi i}{3}$$

D_3	E	$2C_3$	$3C_2'$		
A_1	1	1	1	z^2, x^2+y^2	
A_2	1	1	-1	z	R_z
E	2	-1	0	$\left.\begin{matrix} (xz, yz), (x, y) \\ (x^2-y^2, xy) \end{matrix}\right\}$	(R_x, R_y)

D_{3h}	E	σ_h	$2C_3$	$2S_3$	$3C_2'$	$3\sigma_v$		
A_1'	1	1	1	1	1	1	z^2, x^2+y^2	
A_1''	1	-1	1	-1	1	-1		
A_2'	1	1	1	1	-1	-1		R_z
A_2''	1	-1	1	-1	-1	1	z	
E'	2	2	-1	-1	0	0	$(x, y), (xy, x^2-y^2)$	
E''	2	-2	-1	1	0	0	(xz, yz)	(R_x, R_y)

S_6	E	C_3	C_3^2	i	S_6	S_6^5		
A_g	1	1	1	1	1	1	z^2, x^2+y^2	R_z
A_u	1	1	1	-1	-1	-1	z	
E_g	$\begin{cases}1 \\ 1\end{cases}$ $\begin{matrix}\omega \\ \omega^2\end{matrix}$ $\begin{matrix}\omega^2 \\ \omega\end{matrix}$ $\begin{matrix}1 \\ 1\end{matrix}$ $\begin{matrix}\omega^2 \\ \omega\end{matrix}$ $\begin{matrix}\omega \\ \omega^2\end{matrix}$						$(xz, yz), (x^2-y^2, xy)$	(R_x, R_y)
E_u	$\begin{cases}1 \\ 1\end{cases}$ $\begin{matrix}\omega \\ \omega^2\end{matrix}$ $\begin{matrix}\omega^2 \\ \omega\end{matrix}$ $\begin{matrix}-1 \\ -1\end{matrix}$ $\begin{matrix}-\omega^2 \\ -\omega\end{matrix}$ $\begin{matrix}-\omega \\ -\omega^2\end{matrix}$						(x, y)	

$$\omega = \exp\frac{2\pi i}{3}$$

$D_{3d}(S_{6v})$	E	$2C_3$	$3C_2'$	i	$2S_6$	σ_d		
A_{1g}	1	1	1	1	1	1	z^2, x^2+y^2	
A_{1u}	1	1	1	-1	-1	-1		
A_{2g}	1	1	-1	1	1	-1		R_z
A_{2u}	1	1	-1	-1	-1	1	z	
E_g	2	-1	0	2	-1	0	$(x^2-y^2, xy), (xz, yz)$	(R_x, R_y)
E_u	2	-1	0	-2	1	0	(x, y)	

Fourfold axis

C_4	E	C_2	C_4	C_4^3		
A	1	1	1	1	z, z^2, x^2+y^2	R_z
B	1	1	-1	-1	xy, x^2-y^2	
E	$\begin{cases}1\\1\end{cases}$	$\begin{matrix}-1\\-1\end{matrix}$	$\begin{matrix}i\\-i\end{matrix}$	$\begin{matrix}-i\\i\end{matrix}$	$\left.\begin{matrix}(xz, yz)\\(x, y)\end{matrix}\right\}$	(R_x, R_y)

C_{4v}	E	C_2	$2C_4$	$2\sigma_v$	$2\sigma_v'(2\sigma_d)$		
A_1	1	1	1	1	1	z, z^2, x^2+y^2	
A_2	1	1	1	-1	-1		R_z
B_1	1	1	-1	1	-1	x^2-y^2	
B_2	1	1	-1	-1	1	xy	
E	2	-2	0	0	0	$\left.\begin{matrix}(x, y)\\(xz, yz)\end{matrix}\right\}$	(R_x, R_y)

C_{4h}	E	C_2	C_4	C_4^3	i	σ_h	S_4^3	S_4		
A_g	1	1	1	1	1	1	1	1	z^2, x^2+y^2	R_z
A_u	1	1	1	1	-1	-1	-1	-1	z	
B_g	1	1	-1	-1	1	1	-1	-1	xy, x^2-y^2	
B_u	1	1	-1	-1	-1	-1	1	1		
E_g	$\begin{cases}1\\1\end{cases}$	$\begin{matrix}-1\\-1\end{matrix}$	$\begin{matrix}i\\-i\end{matrix}$	$\begin{matrix}-i\\i\end{matrix}$	$\begin{matrix}1\\1\end{matrix}$	$\begin{matrix}-1\\-1\end{matrix}$	$\begin{matrix}i\\-i\end{matrix}$	$\begin{matrix}-i\\i\end{matrix}$	$\left.\right\}(xz, yz)$	(R_x, R_y)
E_u	$\begin{cases}1\\1\end{cases}$	$\begin{matrix}-1\\-1\end{matrix}$	$\begin{matrix}i\\-i\end{matrix}$	$\begin{matrix}-i\\i\end{matrix}$	$\begin{matrix}-1\\-1\end{matrix}$	$\begin{matrix}1\\1\end{matrix}$	$\begin{matrix}-i\\i\end{matrix}$	$\begin{matrix}i\\-i\end{matrix}$	$\left.\right\}(x, y)$	

D_4	E	C_2	$2C_4$	$2C_2'$	$2C_2''$		
A_1	1	1	1	1	1	z^2, x^2+y^2	
A_2	1	1	1	-1	-1	z	R_z
B_1	1	1	-1	1	-1	x^2-y^2	
B_2	1	1	-1	-1	1	xy	
E	2	-2	0	0	0	$(x, y), (xz, yz)$	(R_x, R_y)

D_{4h}	E	C_2	$2C_4$	$2C_2'$	$2C_2''$	i	σ_h	$2S_4$	$2\sigma_v = 2iC_2'$	$2\sigma_v' = 2iC_2''$		
A_{1g}	1	1	1	1	1	1	1	1	1	1		z^2, x^2+y^2
A_{1u}	1	1	1	1	1	-1	-1	-1	-1	-1		
A_{2g}	1	1	1	-1	-1	1	1	1	-1	-1	R_z	
A_{2u}	1	1	1	-1	-1	-1	-1	-1	1	1		z
B_{1g}	1	1	-1	1	-1	1	1	-1	1	-1		x^2-y^2
B_{1u}	1	1	-1	1	-1	-1	-1	1	-1	1		
B_{2g}	1	1	-1	-1	1	1	1	-1	-1	1		xy
B_{2u}	1	1	-1	-1	1	-1	-1	1	1	-1		
E_g	2	-2	0	0	0	2	-2	0	0	0	(R_x, R_y)	(xz, yz)
E_u	2	-2	0	0	0	-2	2	0	0	0		(x, y)

S_8	E	S_8	C_4	S_8^3	C_2	S_8^5	C_4^3	S_8^7		
A	1	1	1	1	1	1	1	1	z^2, x^2+y^2	R_z
B	1	-1	1	-1	1	-1	1	-1	z	
$E_1 \begin{cases} \\ \end{cases}$	1	ω	ω^2	ω^3	-1	ω^5	ω^6	ω^7	(x, y)	(R_x, R_y)
	1	ω^7	ω^6	ω^5	-1	ω^3	ω^2	ω		
$E_2 \begin{cases} \\ \end{cases}$	1	ω^2	-1	ω^6	1	ω^2	-1	ω^6	(x^2-y^2, xy)	
	1	ω^6	-1	ω^2	1	ω^6	-1	ω^2		
$E_3 \begin{cases} \\ \end{cases}$	1	ω^3	ω^6	ω	-1	ω^7	ω^2	ω^5	(xz, yz)	
	1	ω^5	ω^2	ω^7	-1	ω	ω^6	ω^3		

$$\omega = \exp\frac{2\pi i}{8}$$

$D_{4d}(S_{8v})$	E	C_2	$2C_4$	$2S_8$	$2S_8^3$	$4C_2'$	$4\sigma_d$		
A_1	1	1	1	1	1	1	1	z^2, x^2+y^2	
A_2	1	1	1	1	1	-1	-1		R_z
B_1	1	1	1	-1	-1	1	-1		
B_2	1	1	1	-1	-1	-1	1	z	
E_1	2	-2	0	$\sqrt{2}$	$-\sqrt{2}$	0	0	(x, y)	
E_2	2	2	-2	0	0	0	0	(x^2-y^2, xy)	
E_3	2	-2	0	$-\sqrt{2}$	$\sqrt{2}$	0	0	(xz, yz)	(R_x, R_y)

Fivefold axis

C_5	E	C_5	C_5^2	C_5^3	C_5^4		
A	1	1	1	1	1	z, x^2+y^2, z^2	R_z
E_1	$\begin{cases} 1 \\ 1 \end{cases}$	$\begin{matrix} \omega \\ \omega^4 \end{matrix}$	$\begin{matrix} \omega^2 \\ \omega^3 \end{matrix}$	$\begin{matrix} \omega^3 \\ \omega^2 \end{matrix}$	$\left.\begin{matrix} \omega^4 \\ \omega \end{matrix}\right\}$	$(x, y), (xz, yz)$	(R_x, R_y)
E_2	$\begin{cases} 1 \\ 1 \end{cases}$	$\begin{matrix} \omega^2 \\ \omega^3 \end{matrix}$	$\begin{matrix} \omega^4 \\ \omega \end{matrix}$	$\begin{matrix} \omega \\ \omega^4 \end{matrix}$	$\left.\begin{matrix} \omega^3 \\ \omega^2 \end{matrix}\right\}$	(x^2-y^2, xy)	

$$\omega = \exp\frac{2\pi i}{5}$$

C_{5v}	E	$2C_5$	$2C_5^2$	$5\sigma_v$		
A_1	1	1	1	1	z, z^2, x^2+y^2	
A_2	1	1	1	-1		R_z
E_1	2	$2\cos 72°$	$2\cos 144°$	0	$(x, y), (xz, yz)$	(R_x, R_y)
E_2	2	$2\cos 144°$	$2\cos 288°$	0	(x^2-y^2, xy)	

C_{5h}	E	C_5	C_5^2	C_5^3	C_5^4	σ_h	S_5	$\sigma_h C_5^2 \equiv S_5^7$	S_5^3	$\sigma_h C_5^4 \equiv S_5^9$		
A'	1	1	1	1	1	1	1	1	1	1	R_z	z^2, x^2+y^2
A''	1	1	1	1	1	-1	-1	-1	-1	-1		z
E_1'	$\begin{cases}1\\1\end{cases}$	$\begin{array}{c}\omega\\\omega^4\end{array}$	$\begin{array}{c}\omega^2\\\omega^3\end{array}$	$\begin{array}{c}\omega^3\\\omega^2\end{array}$	$\begin{array}{c}\omega^4\\\omega\end{array}$	$\begin{array}{c}1\\1\end{array}$	$\begin{array}{c}\omega\\\omega^4\end{array}$	$\begin{array}{c}\omega^2\\\omega^3\end{array}$	$\begin{array}{c}\omega^3\\\omega^2\end{array}$	$\begin{array}{c}\omega^4\\\omega\end{array}$		(x, y)
E_1''	$\begin{cases}1\\1\end{cases}$	$\begin{array}{c}\omega\\\omega^4\end{array}$	$\begin{array}{c}\omega^2\\\omega^3\end{array}$	$\begin{array}{c}\omega^3\\\omega^2\end{array}$	$\begin{array}{c}\omega^4\\\omega\end{array}$	$\begin{array}{c}-1\\-1\end{array}$	$\begin{array}{c}-\omega\\-\omega^4\end{array}$	$\begin{array}{c}-\omega^2\\-\omega^3\end{array}$	$\begin{array}{c}-\omega^3\\-\omega^2\end{array}$	$\begin{array}{c}-\omega^4\\-\omega\end{array}$	(R_x, R_y)	(xz, yz)
E_2'	$\begin{cases}1\\1\end{cases}$	$\begin{array}{c}\omega^2\\\omega^3\end{array}$	$\begin{array}{c}\omega^4\\\omega\end{array}$	$\begin{array}{c}\omega\\\omega^4\end{array}$	$\begin{array}{c}\omega^3\\\omega^2\end{array}$	$\begin{array}{c}1\\1\end{array}$	$\begin{array}{c}\omega^2\\\omega^3\end{array}$	$\begin{array}{c}\omega^4\\\omega\end{array}$	$\begin{array}{c}\omega\\\omega^4\end{array}$	$\begin{array}{c}\omega^3\\\omega^2\end{array}$		(x^2, y^2, xy)
E_2''	$\begin{cases}1\\1\end{cases}$	$\begin{array}{c}\omega^2\\\omega^3\end{array}$	$\begin{array}{c}\omega^4\\\omega\end{array}$	$\begin{array}{c}\omega\\\omega^4\end{array}$	$\begin{array}{c}\omega^3\\\omega^2\end{array}$	$\begin{array}{c}-1\\-1\end{array}$	$\begin{array}{c}-\omega^2\\-\omega^3\end{array}$	$\begin{array}{c}-\omega^4\\-\omega\end{array}$	$\begin{array}{c}-\omega\\-\omega^4\end{array}$	$\begin{array}{c}-\omega^3\\-\omega^2\end{array}$		

$$\omega = \exp\frac{2\pi i}{5}$$

D_5	E	$2C_5$	$2C_5^2$	$5C_2'$			
A_1	1	1	1	1		z^2, x^2+y^2	
A_2	1	1	1	-1	z		R_z
E_1	2	$2\cos 72°$	$2\cos 144°$	0		$(x, y), (xz, yz)$	(R_x, R_y)
E_2	2	$2\cos 144°$	$2\cos 288°$	0		(x^2-y^2, xy)	

D_{5h}	E	$2C_5$	$2C_5^2$	$5C_2'$	σ_h	$2S_5$	$2S_5^3$	$5\sigma_v$			
A_1'	1	1	1	1	1	1	1	1		z^2, x^2+y^2	
A_1''	1	1	1	1	-1	-1	-1	-1			
A_2'	1	1	1	-1	1	1	1	-1			R_z
A_2''	1	1	1	-1	-1	-1	-1	1	z		
E_1'	2	$2\cos 72°$	$2\cos 144°$	0	2	$2\cos 72°$	$2\cos 144°$	0		(x, y)	
E_1''	2	$2\cos 72°$	$2\cos 144°$	0	-2	$-2\cos 72°$	$-2\cos 144°$	0		(xz, yz)	(R_x, R_y)
E_2'	2	$2\cos 144°$	$2\cos 288°$	0	2	$2\cos 144°$	$2\cos 288°$	0		(x^2-y^2, xy)	
E_2''	2	$2\cos 144°$	$2\cos 288°$	0	-2	$-2\cos 144°$	$-2\cos 288°$	0			

S_{10}	E	S_{10}	C_5	S_{10}^3	C_5^2	$i=(S_{10}^5)$	C_5^3	S_{10}^7	C_5^4	S_{10}^9		
A_g	1	1	1	1	1	1	1	1	1	1	z^2, x^2+y^2	R_z
A_u	1	1	1	1	1	-1	-1	-1	-1	-1	z	
E_{1g}	1	ω^3	ω	ω^4	ω^2	1	ω^3	ω	ω^4	ω^2	(xz, yz)	(R_x, R_y)
	1	ω^2	ω^4	ω	ω^3	1	ω^2	ω^4	ω	ω^3		
E_{1u}	1	$-\omega^3$	ω	$-\omega^4$	ω^2	-1	ω^3	$-\omega$	ω^4	$-\omega^2$	(x, y)	
	1	$-\omega^2$	ω^4	$-\omega$	ω^3	-1	ω^2	$-\omega^4$	ω	$-\omega^3$		
E_{2g}	1	ω	ω^2	ω^3	ω^4	1	ω	ω^2	ω^3	ω^4	(x^2-y^2, xy)	
	1	ω^4	ω^3	ω^2	ω	1	ω^4	ω^3	ω^2	ω		
E_{2u}	1	$-\omega$	ω^2	$-\omega^3$	ω^4	-1	ω	$-\omega^2$	ω^3	$-\omega^4$		
	1	$-\omega^4$	ω^3	$-\omega^2$	ω	-1	ω^4	$-\omega^3$	ω^2	$-\omega$		

$$\omega = \exp\frac{2\pi i}{5}$$

$D_{5d}(S_{10v})$	E	$2C_5$	$2C_5^2$	$5C_2$	i	$2S_{10}$	$2S_{10}^3$	$5\sigma_d$		
A_{1g}	1	1	1	1	1	1	1	1		z^2, x^2+y^2
A_{1u}	1	1	1	1	-1	-1	-1	-1		
A_{2g}	1	1	1	-1	1	1	1	-1	R_z	
A_{2u}	1	1	1	-1	-1	-1	-1	1	z	
E_{1g}	2	$2\cos 72°$	$2\cos 144°$	0	2	$2\cos 144°$	$2\cos 72°$	0	(R_x, R_y)	(xz, yz)
E_{1u}	2	$2\cos 72°$	$2\cos 144°$	0	-2	$-2\cos 144°$	$-2\cos 72°$	0	(x, y)	
E_{2g}	2	$2\cos 144°$	$2\cos 288°$	0	2	$2\cos 288°$	$2\cos 144°$	0		(x^2-y^2, xy)
E_{2u}	2	$2\cos 144°$	$2\cos 288°$	0	-2	$-2\cos 288°$	$-2\cos 144°$	0		

Sixfold axis

C_6	E	C_6	C_3	C_2	C_3^2	C_6^5		
A	1	1	1	1	1	1	$z, z^2, x^2 - y^2$	R_z
B	1	-1	1	-1	1	-1		
E_1	$\left\{\begin{array}{l}1\\1\end{array}\right.$	$\begin{array}{l}\omega\\\omega^5\end{array}$	$\begin{array}{l}\omega^2\\\omega^4\end{array}$	$\begin{array}{l}-1\\-1\end{array}$	$\begin{array}{l}\omega^4\\\omega^2\end{array}$	$\left.\begin{array}{l}\omega^5\\\omega\end{array}\right\}$	$\left\{\begin{array}{l}(x, y)\\(xz, yz)\end{array}\right\}$	(R_x, R_y)
E_2	$\left\{\begin{array}{l}1\\1\end{array}\right.$	$\begin{array}{l}\omega^2\\\omega^4\end{array}$	$\begin{array}{l}\omega^4\\\omega^2\end{array}$	$\begin{array}{l}1\\1\end{array}$	$\begin{array}{l}\omega^2\\\omega^4\end{array}$	$\left.\begin{array}{l}\omega^4\\\omega^2\end{array}\right\}$	$(x^2 - y^2, xy)$	

$$\omega = \exp\frac{2\pi i}{6}$$

C_{6v}	E	C_2	$2C_3$	$2C_6$	$3\sigma_v$	$3\sigma_v'$		
A_1	1	1	1	1	1	1	$z, z^2, x^2 + y^2$	
A_2	1	1	1	1	-1	-1		R_z
B_1	1	-1	1	-1	-1	1		
B_2	1	-1	1	-1	1	-1		
E_1	2	-2	-1	1	0	0	$(x, y), (xz, yz)$	(R_x, R_y)
E_2	2	2	-1	-1	0	0	$(x^2 - y^2, xy)$	

C_{6h}	E	C_6	C_3	C_2	C_3^2	C_6^5	σ_h	S_6	S_3	i	S_3^5	S_6^5		
A_g	1	1	1	1	1	1	1	1	1	1	1	1	z^2, x^2+y^2	R_z
A_u	1	1	1	1	1	1	-1	-1	-1	-1	-1	-1	z	
B_g	1	-1	1	-1	1	-1	-1	-1	1	1	1	-1		
B_u	1	-1	1	-1	1	-1	1	1	-1	-1	-1	1		
E_{1g}	$\left\{\begin{matrix}1\\1\end{matrix}\right.$	$\begin{matrix}\omega\\\omega^5\end{matrix}$	$\begin{matrix}\omega^2\\\omega^4\end{matrix}$	$\begin{matrix}-1\\-1\end{matrix}$	$\begin{matrix}\omega^4\\\omega^2\end{matrix}$	$\begin{matrix}\omega^5\\\omega\end{matrix}$	$\begin{matrix}-1\\-1\end{matrix}$	$\begin{matrix}\omega^5\\\omega\end{matrix}$	$\begin{matrix}\omega^4\\\omega^2\end{matrix}$	$\begin{matrix}1\\1\end{matrix}$	$\begin{matrix}\omega^2\\\omega^4\end{matrix}$	$\left.\begin{matrix}\omega\\\omega^5\end{matrix}\right\}$	(xz, yz)	(R_x, R_y)
E_{1u}	$\left\{\begin{matrix}1\\1\end{matrix}\right.$	$\begin{matrix}\omega\\\omega^5\end{matrix}$	$\begin{matrix}\omega^2\\\omega^4\end{matrix}$	$\begin{matrix}-1\\-1\end{matrix}$	$\begin{matrix}\omega^4\\\omega^2\end{matrix}$	$\begin{matrix}\omega^5\\\omega\end{matrix}$	$\begin{matrix}1\\1\end{matrix}$	$\begin{matrix}-\omega^5\\-\omega\end{matrix}$	$\begin{matrix}-\omega^4\\-\omega^2\end{matrix}$	$\begin{matrix}-1\\-1\end{matrix}$	$\begin{matrix}-\omega^2\\-\omega^4\end{matrix}$	$\left.\begin{matrix}-\omega\\-\omega^5\end{matrix}\right\}$	(x, y)	
E_{2g}	$\left\{\begin{matrix}1\\1\end{matrix}\right.$	$\begin{matrix}\omega^2\\\omega^4\end{matrix}$	$\begin{matrix}\omega^4\\\omega^2\end{matrix}$	$\begin{matrix}1\\1\end{matrix}$	$\begin{matrix}\omega^2\\\omega^4\end{matrix}$	$\begin{matrix}\omega^4\\\omega^2\end{matrix}$	$\begin{matrix}1\\1\end{matrix}$	$\begin{matrix}\omega^4\\\omega^2\end{matrix}$	$\begin{matrix}\omega^2\\\omega^4\end{matrix}$	$\begin{matrix}1\\1\end{matrix}$	$\begin{matrix}\omega^4\\\omega^2\end{matrix}$	$\left.\begin{matrix}\omega^2\\\omega^4\end{matrix}\right\}$	(x^2-y^2, xy)	
E_{2u}	$\left\{\begin{matrix}1\\1\end{matrix}\right.$	$\begin{matrix}\omega^2\\\omega^4\end{matrix}$	$\begin{matrix}\omega^4\\\omega^2\end{matrix}$	$\begin{matrix}1\\1\end{matrix}$	$\begin{matrix}\omega^2\\\omega^4\end{matrix}$	$\begin{matrix}\omega^4\\\omega^2\end{matrix}$	$\begin{matrix}-1\\-1\end{matrix}$	$\begin{matrix}-\omega^4\\-\omega^2\end{matrix}$	$\begin{matrix}-\omega^2\\-\omega^4\end{matrix}$	$\begin{matrix}-1\\-1\end{matrix}$	$\begin{matrix}-\omega^4\\-\omega^2\end{matrix}$	$\left.\begin{matrix}-\omega^2\\-\omega^4\end{matrix}\right\}$		

$$\omega = \exp\frac{2\pi i}{6}$$

D_6	E	C_2	$2C_3$	$2C_6$	$3C_2'$	$3C_2''$		
A_1	1	1	1	1	1	1		z^2, x^2+y^2
A_2	1	1	1	1	-1	-1	R_z	z
B_1	1	-1	1	-1	1	-1		
B_2	1	-1	1	-1	-1	1		
E_1	2	-2	-1	1	0	0	$(x, y), (R_x, R_y)$	(xz, yz)
E_2	2	2	-1	-1	0	0		(x^2-y^2, xy)

D_{6h}	E	C_2	$2C_3$	$2C_6$	$3C_2'$	$3C_2''$	i	σ_h	$2S_6$	$2S_3$	$3\sigma_v(=iC_2')$	$3\sigma_v'(=iC_2'')$		
A_{1g}	1	1	1	1	1	1	1	1	1	1	1	1		z^2, x^2+y^2
A_{1u}	1	1	1	1	1	1	-1	-1	-1	-1	-1	-1		
A_{2g}	1	1	1	1	-1	-1	1	1	1	1	-1	-1	R_z	
A_{2u}	1	1	1	1	-1	-1	-1	-1	-1	-1	1	1	z	
B_{1g}	1	-1	1	-1	1	-1	1	-1	1	-1	1	-1		
B_{1u}	1	-1	1	-1	1	-1	-1	1	-1	1	-1	1		
B_{2g}	1	-1	1	-1	-1	1	1	-1	1	-1	-1	1		
B_{2u}	1	-1	1	-1	-1	1	-1	1	-1	1	1	-1		
E_{1g}	2	-2	-1	1	0	0	2	-2	-1	1	0	0	(R_x, R_y)	(xz, yz)
E_{1u}	2	-2	-1	1	0	0	-2	2	1	-1	0	0	(x, y)	
E_{2g}	2	2	-1	-1	0	0	2	2	-1	-1	0	0		(x^2-y^2, xy)
E_{2u}	2	2	-1	-1	0	0	-2	-2	1	1	0	0		

S_{12}	E	S_{12}	C_6	S_4	C_3	S_{12}^5	C_2	S_{12}^7	C_3^2	S_4^3	C_6^5	S_{12}^{11}		
A	1	1	1	1	1	1	1	1	1	1	1	1	$z^2,\,x^2+y^2$	R_z
B	1	-1	1	-1	1	-1	1	-1	1	-1	1	-1	z	
$E_1\Big\{$	1	ω	ω^2	ω^3	ω^4	ω^5	-1	ω^7	ω^8	ω^9	ω^{10}	ω^{11}	$(x,y),$	
	1	ω^{11}	ω^{10}	ω^9	ω^8	ω^7	-1	ω^5	ω^4	ω^3	ω^2	ω		
$E_2\Big\{$	1	ω^2	ω^4	-1	ω^8	ω^{10}	1	ω^2	ω^4	-1	ω^8	ω^{10}	$(x^2-y^2,\,xy)$	
	1	ω^{10}	ω^8	-1	ω^4	ω^2	1	ω^{10}	ω^8	-1	ω^4	ω^2		
$E_3\Big\{$	1	ω^3	-1	ω^9	1	ω^3	-1	ω^9	1	ω^3	-1	ω^9		
	1	ω^9	-1	ω^3	1	ω^9	-1	ω^3	1	ω^9	-1	ω^3		
$E_4\Big\{$	1	ω^4	ω^8	1	ω^4	ω^8	1	ω^4	ω^8	1	ω^4	ω^8		
	1	ω^8	ω^4	1	ω^8	ω^4	1	ω^8	ω^4	1	ω^8	ω^4		
$E_5\Big\{$	1	ω^5	ω^{10}	ω^3	ω^8	ω	-1	ω^{11}	ω^4	ω^9	ω^2	ω^7	$(xz,\,yz)$	$(R_x,\,R_y)$
	1	ω^7	ω^2	ω^9	ω^4	ω^{11}	-1	ω	ω^8	ω^3	ω^{10}	ω^5		

$$\omega = \exp\frac{2\pi i}{12}$$

$D_{6d}(S_{12v})$	E	$2S_{12}$	$2C_6$	$2S_4$	$2C_3$	$2S_{12}^5$	C_2	$6C_2'$	$6\sigma_d$		
A_1	1	1	1	1	1	1	1	1	1		z^2, x^2+y^2
A_2	1	1	1	1	1	1	1	-1	-1	R_z	
B_1	1	-1	1	-1	1	-1	1	1	-1		
B_2	1	-1	1	-1	1	-1	1	-1	1		z
E_1	2	$\sqrt{3}$	1	0	-1	$-\sqrt{3}$	-2	0	0		(x,y),
E_2	2	1	-1	-2	-1	1	2	0	0		(x^2-y^2, xy)
E_3	2	0	-2	0	2	0	-2	0	0		
E_4	2	-1	-1	2	-1	-1	2	0	0		
E_5	2	$-\sqrt{3}$	1	0	-1	$\sqrt{3}$	-2	0	0	(R_x, R_y)	(xz, yz)

Continuous groups

C_∞	E	$\overrightarrow{C_x}$	$\overleftarrow{C_x}$		
$A\ (\Sigma)$	1	1	1	z, z^2, x^2+y^2	R_z
$E_1\ (\Pi)$	$\begin{cases} 1 \\ 1 \end{cases}$	$\begin{matrix} \omega \\ \omega^{-1} \end{matrix}$	$\left.\begin{matrix} \omega^{-1} \\ \omega \end{matrix}\right\}$	$(x, y), (xy, yz)$	(R_x, R_y)
$E_2\ (\Delta)$	$\begin{cases} 1 \\ 1 \end{cases}$	$\begin{matrix} \omega^2 \\ \omega^{-2} \end{matrix}$	$\left.\begin{matrix} \omega^{-2} \\ \omega^2 \end{matrix}\right\}$	(x^2-y^2, xy)	
$\vdots$	$\vdots$	$\vdots$	$\vdots$		
E_k	$\begin{cases} 1 \\ 1 \end{cases}$	$\begin{matrix} \omega^k \\ \omega^{-k} \end{matrix}$	$\begin{matrix} \omega^{-k} \\ \omega^k \end{matrix}$		
$\vdots$	$\vdots$	$\vdots$	$\vdots$		

$$x = \frac{2\pi}{\theta} \qquad \omega = \exp\frac{2\pi i}{\theta}$$

$C_{\infty v}$	E	$2C_x$	σ_v		
$A_1\ (\Sigma^+)$	1	1	1	z, z^2, x^2+y^2	
$A_2\ (\Sigma^-)$	1	1	-1		R_z
$E_1\ (\Pi)$	2	$2\cos\theta$	0	$(x, y), (xz, yz)$	(R_x, R_y)
$E_2\ (\Delta)$	2	$2\cos 2\theta$	0	(x^2-y^2, xy)	
$\vdots$	$\vdots$	$\vdots$	$\vdots$		
E_k	2	$2\cos k\theta$	0		
$\vdots$	$\vdots$	$\vdots$	$\vdots$		

$$x = \frac{2\pi}{\theta}$$

D_∞	E	$2C_x$	C_2'		
$A_1\ (\Sigma^+)$	1	1	1	z^2, x^2+y^2	
$A_2\ (\Sigma^-)$	1	1	-1	z	R_z
$E_1\ (\Pi)$	2	$2\cos\theta$	0	$(x, y), (xz, yz)$	(R_x, R_y)
$E_2\ (\Delta)$	2	$2\cos 2\theta$	0	(x^2-y^2, xy)	
$\vdots$	$\vdots$	$\vdots$	$\vdots$		
E_k	2	$2\cos k\theta$	0		
$\vdots$	$\vdots$	$\vdots$	$\vdots$		

$$x = \frac{2\pi}{\theta}$$

$D_{\infty h}$	E	$2C_x$	$\infty C'_2$	i	$2iC_x$	(σ_h)	$\infty \sigma_v$		
$A_{1g}(\Sigma_g^+)$	1	1	1	1	1	1	1		z^2, x^2+y^2
$A_{1u}(\Sigma_u^-)$	1	1	1	-1	-1	-1	-1		
$A_{2g}(\Sigma_g^-)$	1	1	-1	1	1	1	-1	R_z	
$A_{2u}(\Sigma_u^+)$	1	1	-1	-1	-1	-1	1	z	
$E_{1g}(\Pi_g)$	2	$2\cos\theta$	0	2	$2\cos\theta$	-2	0	(R_x, R_y)	(xz, yz)
$E_{1u}(\Pi_u)$	2	$2\cos\theta$	0	-2	$-2\cos\theta$	2	0		(x, y)
$E_{2g}(\Delta_g)$	2	$2\cos 2\theta$	0	2	$2\cos 2\theta$	2	0		(x^2-y^2, xy)
$E_{2u}(\Delta_u)$	2	$2\cos 2\theta$	0	-2	$-2\cos 2\theta$	-2	0		
...	...	...	...	...	...	...	...		
E_{kg}	2	$2\cos k\theta$	0	2	$2\cos k\theta$	$2(-1)^k$	0		
E_{ku}	2	$2\cos k\theta$	0	-2	$-2\cos k\theta$	$2(-1)^{k+1}$	0		
...	...	...	...	...	...	...	...		

$x = \dfrac{2\pi}{\theta}$ σ_h is a special case of $2iC_x$, $x = 2$

Tetrahedral groups

T	E	$3C_2$	$4C_3$	$4C_3^2$		
A	1	1	1	1	$(x^2+y^2+z^2)$	
E	1	1	ω	ω^2	$(x^2-y^2),\,(2z^2-x^2+y^2)$	
	1	1	ω^2	ω		
T	3	-1	0	0	(x,y,z); (xy,xz,yz)	(R_x, R_y, R_z)

$$\omega = \exp\frac{2\pi i}{3}$$

S_{10}	E	S_{10}	C_5	$(S_{10})^3$	$(C_5)^2$	$(S_{10})^5$	$(C_5)^3$	$(S_{10})^7$	$(C_5)^4$	$(S_{10})^9$		
A_g	1	1	1	1	1	1	1	1	1	1	z^2, x^2+y^2	R_z
A_u	1	-1	1	-1	1	-1	1	-1	1	-1	z	
E_{1g}	1	w^3	w	w^4	w^2	1	w^3	w	w^4	w^2	(xz, yz)	(R_x, R_y)
	1	w^2	w^4	w	w^3	1	w^2	w^4	w	w^3		
E_{1u}	1	$-w^3$	w	$-w^4$	w^2	-1	w^3	$-w$	w^4	$-w^2$	(x, y)	
	1	$-w^2$	w^4	$-w$	w^3	-1	w^2	$-w^4$	w	$-w^3$		
E_{2g}	1	w	w^2	w^3	w^4	1	w	w^2	w^3	w^4	(x^2-y^2, xy)	
	1	w^4	w^3	w^2	w	1	w^4	w^3	w^2	w		
E_{2u}	1	$-w$	w^2	$-w^3$	w^4	-1	w	$-w^2$	w^3	$-w^4$		
	1	$-w^4$	w^3	$-w^2$	w	-1	w^4	$-w^3$	w^2	$-w$		

D_5 (S_{10r})	E	$2C_5$	$2(C_5)^2$	$5C_2'$	i	$2S_{10}$	$2(S_{10})^3$	$5\sigma_d$		
A_{1g}	1	1	1	1	1	1	1	1		z^2, x^2+y^2
A_{1u}	1	1	1	1	-1	-1	-1	-1		
A_{2g}	1	1	1	-1	1	1	1	-1	R_z	
A_{2u}	1	1	1	-1	-1	-1	-1	1	z	
E_{1g}	2	$2\cos 72°$	$2\cos 144°$	0	2	$2\cos 144$	$2\cos 72°$	0		(xz, yz)
E_{1u}	2	$2\cos 72°$	$2\cos 144$	0	-2	$-2\cos 144$	$-2\cos 72°$	0	(R_x, R_y)	(x, y)
E_{2g}	2	$2\cos 144°$	$2\cos 288°$	0	2	$2\cos 288°$	$2\cos 144°$	0		$(x-y, xy)$
E_{2u}	2	$2\cos 144°$	$2\cos 288°$	0	2	$-2\cos 288°$	$-2\cos 144°$	0		

T_h	E	$3C_2$	$4C_3$	$4C_3^2$	i	$3\sigma_h(=iC_2)$	$4S_6$	$4S_6^5$	
A_g	1	1	1	1	1	1	1	1	$(x^2+y^2+z^2)$
A_u	1	1	1	1	-1	-1	-1	-1	
E_g	$\left\{\begin{array}{l}1\\1\end{array}\right.$	1 1	$\begin{array}{l}\omega\\\omega^2\end{array}$	$\begin{array}{l}\omega^2\\\omega\end{array}$	1 1	1 1	$\left.\begin{array}{l}\omega\\\omega^2\end{array}\right\}$	$\begin{array}{l}\omega^2\\\omega\end{array}$	$(x^2-y^2),\,(2z^2-x^2+y^2)$
E_u	$\left\{\begin{array}{l}1\\1\end{array}\right.$	1 1	$\begin{array}{l}\omega\\\omega^2\end{array}$	$\begin{array}{l}\omega^2\\\omega\end{array}$	$\begin{array}{l}-1\\-1\end{array}$	$\begin{array}{l}-1\\-1\end{array}$	$\left.\begin{array}{l}-\omega\\-\omega^2\end{array}\right\}$	$\begin{array}{l}-\omega^2\\-\omega\end{array}$	
T_g	3	-1	0	0	3	-1	0	0	(xy, xz, yz) (R_x, R_y, R_z)
T_u	3	-1	0	0	-3	1	0	0	(x, y, z)

$$\omega = \exp\frac{2\pi i}{3}$$

T_d	E	$8C_3$	$3C_2$	$6\sigma_d$	$6S_4$	
A_1	1	1	1	1	1	$x^2+y^2+z^2$
A_2	1	1	1	-1	-1	
E	2	-1	2	0	0	$(x^2-y^2,\,2z^2-(x^2+y^2))$
T_1	3	0	-1	-1	1	(R_x, R_y, R_z)
T_2	3	0	-1	1	-1	$\left\{\begin{array}{l}(x,y,z)\\(xy, xz, yz)\end{array}\right\}$

Octahedral groups

O	E	$8C_3$	$3C_2(=C_4^2)$	$6C_2'$	$6C_4$		
A_1	1	1	1	1	1	$x^2+y^2+z^2$	
A_2	1	1	1	-1	-1		
E	2	-1	2	0	0	$(x^2-y^2, 2z^2-(x^2+y^2))$	
T_1	3	0	-1	-1	1	(R_x, R_y, R_z) (x, y, z)	
T_2	3	0	-1	1	-1	(xy, xz, yz)	

O_h	E	$8C_3$	$3C_2$	$6C_2'$	$6C_4$	i	$8S_6$	$3\sigma_h$	$6\sigma_v$	$6S_4$	
A_{1g}	1	1	1	1	1	1	1	1	1	1	$(x^2+y^2+z^2)$
A_{1u}	1	1	1	1	1	-1	-1	-1	-1	-1	
A_{2g}	1	1	1	-1	-1	1	1	1	-1	-1	
A_{2u}	1	1	1	-1	-1	-1	-1	-1	1	1	
E_g	2	-1	2	0	0	2	-1	2	0	0	$(x^2-y^2, 2z^2-(x^2+y^2))$
E_u	2	-1	2	0	0	-2	1	-2	0	0	
T_{1g}	3	0	-1	-1	1	3	0	-1	-1	1	(R_x, R_y, R_z)
T_{1u}	3	0	-1	-1	1	-3	0	1	1	-1	(x, y, z)
T_{2g}	3	0	-1	1	-1	3	0	-1	1	-1	(xy, xz, yz)
T_{2u}	3	0	-1	1	-1	-3	0	1	-1	1	

Icosahedral groups

P	E	$15C_2$	$20C_3$	$12C_5$	$12C_5^2$		
A	1	1	1	1	1	$x^2+y^2+z^2$	
T_1	3	-1	0	m	n	(x, y, z)	(R_x, R_y, R_z)
T_2	3	-1	0	n	m		
U	4	0	1	-1	-1		
V	5	1	-1	0	0	$\left\{\begin{array}{l}(xy, xz, yz, x^2-y^2, \\ 2z^2-(x^2+y^2))\end{array}\right\}$	

$$m = 1+2\cos 72° = \tfrac{1}{2}(\sqrt{5}+1) = +1{\cdot}618$$
$$n = 1+2\cos 144° = -\tfrac{1}{2}(\sqrt{5}-1) = -0{\cdot}618$$

$P_h \equiv P_i$	E	$15C_2$	$20C_3$	$12C_5$	$12C_5^2$	i	$15\sigma_h(=iC_2)$	$20S_6$	$12S_{10}^3$	$12S_{10}$		
A_g	1	1	1	1	1	1	1	1	1	1	$(x^2+y^2+z^2)$	
A_u	1	1	1	1	1	-1	-1	-1	-1	-1		
T_{1g}	3	-1	0	m	n	3	-1	0	m	n		(R_x, R_y, R_z)
T_{1u}	3	-1	0	m	n	-3	1	0	$-m$	$-n$	(x, y, z)	
T_{2g}	3	-1	0	n	m	3	-1	0	n	m		
T_{2u}	3	-1	0	n	m	-3	1	0	$-n$	$-m$		
U_g	4	0	1	-1	-1	4	0	1	-1	-1		
U_u	4	0	1	-1	-1	-4	0	-1	1	1		
V_g	5	1	-1	0	0	5	1	-1	0	0	$(xy, xz, yz, x^2-y^2, 2z^2-(x^2+y^2))$	
V_u	5	1	-1	0	0	-5	-1	1	0	0		

$$m = 1+2\cos 72° = \tfrac{1}{2}(\sqrt{5}+1) = +1{\cdot}618$$
$$n = 1+2\cos 144° = -\tfrac{1}{2}(\sqrt{5}-1) = -0{\cdot}618$$

Answers

1.1 If $\psi = A \sin Bx$,

then $\dfrac{\partial^2 \psi}{\partial x^2} = -B^2 \psi$

and ψ is an eigenfunction,

but $\dfrac{\partial^2 \psi}{\partial x^2} = -\dfrac{8\pi^2 mE}{h^2} \psi$,

thus $E = \dfrac{h^2 B^2}{8\pi^2 m}$.

If $\psi = A \sin Bx$ then $\psi = 0$ for $x = 0$.
If ψ is to be 0 when $x = a$, then Ba must be π or an integral multiple of π,

i.e. $Ba = n\pi \quad (n = 1, 2, 3, \ldots)$.

Thus $E = \dfrac{n^2 h^2}{8ma^2}$.

Normalization:

$$1 = A^2 \int_0^a \sin^2 Bx \, dx$$

$$= A^2 \int_0^a \tfrac{1}{2}(1 - \cos 2Bx) \, dx$$

$$= \frac{A^2}{2}\left[x - \frac{\sin 2Bx}{2B} \right]_0^a = \frac{A^2 a}{2}.$$

Hence $A = \sqrt{\dfrac{2}{a}}$

and the eigenfunction is

$$\psi = \sqrt{\left(\frac{2}{a}\right)} \sin \frac{n\pi x}{a}.$$

The eigenvalues in the 'butadiene' box will be:

$$E = \frac{n^2 \times 6.6 \times 6.6 \times 10^{-54}}{8 \times 9 \times 10^{-28} \times 4.2 \times 4.2 \times 10^{-16}} \times \frac{1}{1.6 \times 10^{-12}} \text{ eV}$$

$$= 2.1n^2 \text{ eV}.$$

The first three eigenvalues are 2.1, 8.4 and 18.9 eV, and so the energy difference between the second and the third is 10.5 eV. This is really remarkably good agreement with the behaviour of a charged electron in a complex potential field.

1.2 If $x = r\phi$ and r is a constant, the Schrödinger equation is

$$-\frac{h^2}{8\pi^2 m r^2} \frac{\partial^2 \psi}{\partial \phi^2} = E\psi,$$

$$\frac{\partial^2 \psi}{\partial \phi^2} = -D^2 \psi \quad \text{and} \quad \frac{\partial^2 \psi'}{\partial \phi^2} = -D'^2 \psi',$$

thus both ψ and ψ' are eigenfunctions.

The 'boundary condition' requires that both D and D' be integral with the following very important difference:

$$D = 1, 2, 3, \ldots$$

but $D' = 0, 1, 2, 3, \ldots$

This is because when $D = 0, \psi = 0$ but when $D' = 0$ the function does not vanish (it just becomes independent of ϕ).

Normalization:

$$1 = C^2 \int_0^{2\pi} \sin^2 D\phi \, d\phi$$

$$= \frac{C^2}{2} \left[\phi - \frac{\sin 2D\phi}{2D} \right]_0^{2\pi} = \pi C^2.$$

Thus $C = \dfrac{1}{\sqrt{\pi}}.$

Also $C' = \dfrac{1}{\sqrt{\pi}} \quad (D' \neq 0)$

when $D' = 0, C' = \dfrac{1}{\sqrt{(2\pi)}}.$

General orthogonality: consider $\int_0^{2\pi} \sin n\phi \cos m\phi \, d\phi$, where n and m can be any integers.

$$\int_0^{2\pi} \sin n\phi \cos m\phi \, d\phi = \tfrac{1}{2} \int_0^{2\pi} \sin(n+m)\phi + \sin(n-m)\phi \, d\phi$$

$$= \frac{1}{2}\left[-\frac{\cos(n+m)\phi}{(n+m)} - \frac{\cos(n-m)\phi}{(n-m)} \right]_0^{2\pi}$$

$$= 0.$$

Thus any eigenfunction is orthogonal to any other eigenfunction.
Energies:

$$\psi = \frac{1}{\sqrt{\pi}} \sin D\phi \qquad (D = 1, 2, 3, \ldots),$$

$$\psi' = \frac{1}{\sqrt{\pi}} \cos D'\phi \qquad (D' = 1, 2, 3, \ldots),$$

$$\text{also} \quad \psi = \frac{1}{\sqrt{(2\pi)}} \qquad (D' = 0).$$

For both functions $\quad E = \dfrac{h^2(D \text{ or } D')^2}{8\pi^2 mr^2}$

$$= 1.93 \times (D \text{ or } D')^2 \text{ eV}.$$

The energies are therefore 0, 1.93 and 7.72 eV (for ψ') and 1.93 and 7.72 eV (for ψ).

The difference between first and second eigenvalues is 5.79 eV in very close agreement with the average energy of the first three u.v. absorption bands of benzene (~ 6 eV).

(When $D' = 0$, the total energy is found to be zero. Since $V = 0$ has been assumed this means the kinetic energy and therefore the momentum are also zero. If, however, the momentum is known exactly the position of the particle must lack definition by the uncertainty principle: the wavefunction $\psi' = (2\pi)^{-\frac{1}{2}}$ is in accord with this conclusion since it is independent of ϕ.)

1.3 *To show*
$$+1 = \tfrac{1}{2}\!\int\!\int 1s(1)1s(2)\{\alpha(1)\beta(2) - \beta(1)\alpha(2)\}1s(1)1s(2)\{\alpha(1)\beta(2) - \beta(1)\alpha(2)\} \, d\tau(1) \, d\tau(2).$$
The integral may be expanded into orbital (dv) and spin (ds) parts for the two electrons, ($d\tau = dv \times ds$):

$$\tfrac{1}{2}\{\int 1s(1)1s(1)\,dv(1) \int 1s(2)1s(2)\,dv(2) \times$$
$$\times [\int \alpha(1)\alpha(1)\,ds(1) \int \beta(2)\beta(2)\,ds(2) - \int \alpha(1)\beta(1)\,ds(1) \int \beta(2)\alpha(2)\,ds(2) -$$
$$- \int \beta(1)\alpha(1)\,ds(1) \int \alpha(2)\beta(2)\,ds(2) + \int \beta(1)\beta(1)\,ds(1) \int \alpha(2)\alpha(2)\,ds(2)]\}.$$

Writing down the value of each integral, this becomes
$$\tfrac{1}{2}\{1 \times 1[1 \times 1 - 0 \times 0 - 0 \times 0 + 1 \times 1]\} = \tfrac{1}{2}.1.2$$

$$= +1.$$

1.4 $\quad \psi_I = \frac{1}{2}\left\{\begin{vmatrix} 1s\alpha(1) & 2s\beta(1) \\ 1s\alpha(2) & 2s\beta(2) \end{vmatrix} - \begin{vmatrix} 1s\beta(1) & 2s\alpha(1) \\ 1s\beta(2) & 2s\alpha(2) \end{vmatrix}\right\},$

$\quad\psi_{II} = \frac{1}{2}\left\{\begin{vmatrix} 1s\alpha(1) & 2s\beta(1) \\ 1s\alpha(2) & 2s\beta(2) \end{vmatrix} + \begin{vmatrix} 1s\beta(1) & 2s\alpha(1) \\ 1s\beta(2) & 2s\alpha(2) \end{vmatrix}\right\},$

$\quad\psi_{III} = \frac{1}{\sqrt{2}}\begin{vmatrix} 1s\alpha(1) & 2s\alpha(1) \\ 1s\alpha(2) & 2s\alpha(2) \end{vmatrix},$

$\quad\psi_{IV} = \frac{1}{\sqrt{2}}\begin{vmatrix} 1s\beta(1) & 2s\beta(1) \\ 1s\beta(2) & 2s\beta(2) \end{vmatrix}.$

Notice that the actual wave functions may sometimes be linear combinations of Slater determinants. The allocation of electrons to particular spin orbitals, as in the Slater determinants, defines a particular microstate (see Chapter 6, p. 167).

1.5 The evaluation of the energy corresponding to E_{II} is exactly as for E_I (p. 23) save for those terms which have changed signs.

$E_{III} \int \psi_{III} \mathscr{H} \psi_{III} \, d\tau$

$$= \frac{1}{2}\iint \{1s\alpha(1)2s\alpha(2) - 2s\alpha(1)1s\alpha(2)\}\left[\mathscr{H}(1) + \mathscr{H}(2) + \frac{e^2}{r(12)}\right] \times$$

$$\times \{1s\alpha(1)2s\alpha(2) - 2s\alpha(1)1s\alpha(2)\} \, d\tau(1) \, d\tau(2).$$

Let $\quad 1s\alpha(1)2s\alpha(2) = Q \quad$ and $\quad 2s\alpha(1)1s\alpha(2) = R.$

Then $\quad E_{III} = \frac{1}{2}\iint Q(\ )Q \, d\tau(1)d\tau(2) + \frac{1}{2}\iint R(\ )R \, d\tau(1)d\tau(2) -$
$\qquad\qquad\qquad - \frac{1}{2}\iint Q(\ )R \, d\tau(1)d\tau(2) - \frac{1}{2}\iint R(\ )Q \, d\tau(1)d\tau(2).$

Now

$\frac{1}{2}\iint Q(\ )Q \, d\tau(1)d\tau(2) = \frac{1}{2}\{E_{1s}(1) + E_{2s}(2) + J_{1s2s}\},$

$\frac{1}{2}\iint R(\ )R \, d\tau(1)d\tau(2) = \frac{1}{2}\{E_{2s}(1) + E_{1s}(2) + J_{1s2s}\},$

$\frac{1}{2}\iint Q(\ )R \, d\tau(1)d\tau(2)$

$$= -\frac{1}{2}\iint 1s\alpha(1)2s\alpha(2)\left[\mathscr{H}(1) + \mathscr{H}(2) + \frac{e^2}{r(12)}\right]2s\alpha(1)1s\alpha(2) \, d\tau(1)d\tau(2)$$

$$= -\frac{1}{2}\int 1s\alpha(1)\mathscr{H}(1)2s\alpha(1) \, d\tau(1)\underbrace{\int 2s\alpha(2)1s\alpha(2) \, d\tau(2)}_{\text{zero}} \, -$$

$$-\frac{1}{2}\int 2s\alpha(2)\mathscr{H}(2)1s\alpha(2) \, d\tau(2)\underbrace{\int 1s\alpha(1)2s\alpha(1) \, d\tau(1)}_{\text{zero}} \, -$$

$$-\frac{1}{2}\iint 1s\alpha(1)2s\alpha(2)\left[\frac{e^2}{r(12)}\right]2s\alpha(1)1s\alpha(2) \, d\tau(1)d\tau(2).$$

Since all the spin functions are normalized and identical, the integrals $-\frac{1}{2}\int Q(\ )R\,d\tau - \frac{1}{2}\int R(\ )Q\,d\tau$ simply reduce to $-K_{1s2s}$.

Thus $E_{III} = E_{1s} + E_{2s} + J_{1s2s} - K_{1s2s}$.

E_{IV} can be found in the same way.

Thus $E_{II} = E_{III} = E_{IV}$.

2.1   The determinant is $\begin{vmatrix} x & 1 \\ 1 & x \end{vmatrix} = 0$. Hence $x = \pm 1$. In the bonding orbital ($x = -1$) the m.o. is $\psi(\text{bond}) = (\sqrt{\frac{1}{2}})(\psi_a + \psi_b)$.

For $x = +1$ $\psi(\text{antibond}) = (\sqrt{\frac{1}{2}})(\psi_a - \psi_b)$;
ψ_a and ψ_b are the two hydrogen $1s$ atomic orbitals from the two atoms in H–H. If overlap is included

$$\begin{vmatrix} \alpha - E & \beta - ES \\ \beta - ES & \alpha - E \end{vmatrix} = 0.$$

If $x = (\alpha - E)/(\beta - ES)$ we can use the determinant above. Then, for the bonding orbital

$$\alpha - E = (-1)(\beta - ES),$$

i.e.
$$E = \frac{\alpha + \beta}{1 + S}.$$

In general
$$E = \frac{\alpha - x\beta}{1 - xS}.$$

Thus, for $x = +1$ $E = \frac{\alpha - \beta}{1 - S}.$

Since S will be a small positive fraction (e.g. $\frac{1}{4} - \frac{1}{5}$ for C–C π-bonds) the effect of including overlap will be to make all bonding orbitals less tightly bound and the antibonding orbitals more tightly bound, than the calculations in which S is neglected would suggest.

2.2   For square cyclobutadiene we can assume all the resonance integrals between adjacent p-orbitals to be the same. The determinant for the π molecular orbitals is

$$\begin{vmatrix} x & 1 & 0 & 1 \\ 1 & x & 1 & 0 \\ 0 & 1 & x & 1 \\ 1 & 0 & 1 & x \end{vmatrix} = 0.$$

Hence $x^2(x^2 - 4) = 0$,
i.e. $x = \pm 2, 0, 0.$

To allocate four electrons to these orbitals would mean putting two into the $x = -2$ orbital and then, for the ground state, two electrons with parallel spin one into each of the $x = 0$ orbitals, that is a triplet ground state – a diradical.

2.3 Let the resonance integrals between orbitals 1 and 2 and between 2 and 3 be β, then for bond 1–3 the integral is $k\beta$.

The determinant is

$$\begin{vmatrix} x & 1 & k \\ 1 & x & 1 \\ k & 1 & x \end{vmatrix} = 0.$$

Expanding the determinants,

$$x(x^2 - 1) - (x - k) + k(1 - xk) = 0,$$

rearranging, $\qquad x^3 - 2x + 2k - xk^2 = 0,$

factorizing, $\qquad x(x^2 - k^2) + 2(k - x) = 0,$

$$x(x + k)(x - k) - 2(x - k) = 0,$$

$$(x - k)(x^2 + xk - 2) = 0.$$

Thus $\quad x = +k \quad$ or $\quad x = \frac{1}{2}\{-k \pm \sqrt{(k^2 + 8)}\}.$

Both of these functions can easily be plotted graphically x against k.

3.1 ICl_4^-

C_4, C_2, C_4^3, S_4, S_4^3 all on same axis through I perpendicular to the plane of the ion.

C_2' along both Cl–I–Cl lines and also C_2'' bisecting both Cl–I–Cl right angles.

σ_h.

σ_vs containing z-axis and C_2's.

σ_v's containing z-axis and C_2''s.

Centre of symmetry.

Al_2Cl_6

Three C_2 axes mutually perpendicular meet at the middle of the molecule. Each pair of axes contains a mirror plane. Centre of symmetry.

P_4O_{10}

C_3 and C_3^2 operations share axes along each of the external P–O bonds.

C_2 axes join midpoints of opposite edges (i.e. bridging oxygen atoms).

S_4 axes found at the same sites.

Mirror planes contain two P–O external bonds (six such planes).

Anthracene
As for Al_2Cl_6.

Phenanthrene
A C_2 axis in the plane of the molecule bisects and is at right angles to the 9, 10 C–C bond.

σ_v contains C_2 and is perpendicular to the molecular plane.

σ_v' contains C_2 and is in the molecular plane.

1,3,5-Trifluorobenzene
A threefold axis is perpendicular to the molecule and passes through the centre of the benzene ring. C_3, C_3^2 – also site of S_3 and S_3^5 – C_2' axes pass through C–F and opposite C–H bonds.

σ_h.

σ_v planes contain C_2' and z axes (three such).

Formaldehyde
C_2 axis along the C–O bond.

Mirror planes σ_v are perpendicular to the molecule and the plane of the molecule (both contain C_2).

3.2 Symmetry operations as for Al_2Cl_6. Let the z-axis be perpendicular to the plane of the molecule and let the x-axis join the carbon atoms.

first $\rightarrow$ $C_2(x)$ $C_2(y)$ $C_2(z)$ $\sigma(xy)$ $\sigma(xz)$ $\sigma(yz)$

second
$\downarrow$

$C_2(x)$	E	$C_2(z)$	$C_2(y)$	$\sigma(xz)$	$\sigma(xy)$	i
$C_2(y)$	$C_2(z)$	E	$C_2(x)$	$\sigma(yz)$	i	$\sigma(xy)$
$C_2(z)$	$C_2(y)$	$C_2(x)$	E	i	$\sigma(yz)$	$\sigma(xz)$
$\sigma(xy)$	$\sigma(xz)$	$\sigma(xy)$	i	E	$C_2(x)$	$C_2(y)$
$\sigma(xz)$	$\sigma(yz)$	i	$\sigma(xy)$	$C_2(x)$	E	$C_2(z)$
$\sigma(yz)$	i	$\sigma(yz)$	$\sigma(xz)$	$C_2(y)$	$C_2(z)$	E

The operations do form a group.
Each operation is its own inverse.
All the operations commute.

3.3 ICl_4^-, $\boldsymbol{D_{4h}}$; Al_2Cl_6, $\boldsymbol{D_{2h}}$; P_4O_{10}, $\boldsymbol{T_d}$; anthracene, $\boldsymbol{D_{2h}}$; phenanthrene $\boldsymbol{C_{2v}}$; 1,3,5-trifluorobenzene, $\boldsymbol{D_{3h}}$; formaldehyde, $\boldsymbol{C_{2v}}$.

3.4 (a) $\boldsymbol{O_h}$.
     (b) $\boldsymbol{C_{3v}}$.

4.1 *Operations*   *Test representation*
                          actual test representation   the test representation
                          product   product should equal

$EC_2 \qquad = C_2 \qquad\qquad 1 \times 1 = 1$ $1\ \checkmark$

$E\sigma_v'(yz) \quad = \sigma_v'(yz) \qquad 1 \times 1 = 1$ $1\ \checkmark$

$E\sigma_v(xz) \quad = \sigma_v(xz) \qquad 1 \times -1 = -1$ $-1\ \checkmark$

$C_2\,\sigma_v(xz) \quad = \sigma_v'(yz) \qquad 1 \times -1 = -1$ $1\ \times$

$C_2\,\sigma_v'(yz) \quad = \sigma_v(xz) \qquad 1 \times 1 = 1$ $-1\ \times$

$\sigma_v(xz)\sigma_v'(yz) = C_2 \qquad 1 \times 1 = 1$ $1\ \times$

For the test representation to be a true representation there should be complete
agreement between the two columns. The presence of any $\times$s would mean the
test representation is *not* a true representation.

4.2
$$
\begin{bmatrix} d_{z^2} \\ d_{x^2-y^2} \\ d_{xy} \\ d_{xz} \\ d_{yz} \end{bmatrix},
$$

E
$$
\begin{bmatrix}
1 & 0 & 0 & 0 & 0 \\
0 & 1 & 0 & 0 & 0 \\
0 & 0 & 1 & 0 & 0 \\
0 & 0 & 0 & 1 & 0 \\
0 & 0 & 0 & 0 & 1
\end{bmatrix}
$$

C_2
$$
\begin{bmatrix}
1 & 0 & 0 & 0 & 0 \\
0 & 1 & 0 & 0 & 0 \\
0 & 0 & 1 & 0 & 0 \\
0 & 0 & 0 & -1 & 0 \\
0 & 0 & 0 & 0 & -1
\end{bmatrix}
$$

$\sigma_v(xz)$
$$
\begin{bmatrix}
1 & 0 & 0 & 0 & 0 \\
0 & 1 & 0 & 0 & 0 \\
0 & 0 & -1 & 0 & 0 \\
0 & 0 & 0 & 1 & 0 \\
0 & 0 & 0 & 0 & -1
\end{bmatrix}
$$

$\sigma_v'(yz)$
$$
\begin{bmatrix}
1 & 0 & 0 & 0 & 0 \\
0 & 1 & 0 & 0 & 0 \\
0 & 0 & -1 & 0 & 0 \\
0 & 0 & 0 & -1 & 0 \\
0 & 0 & 0 & 0 & 1
\end{bmatrix}
$$

4.3 $\chi_E = 5$ $\chi_{C_2} = 1$ $\chi_{v(xz)}^\sigma = 1$ $\chi_{v(yz)}^\sigma = 1.$

	E	C_2	$\sigma_v(xz)$	$\sigma_v'(yz)$
$a(A_1) = \frac{1}{4}(1 \times 1 \times 5 + 1 \times 1 \times 1$			$+1 \times 1 \times 1$	$+1 \times 1 \times 1)$ $= 2$
$a(A_2) = \frac{1}{4}(1 \times 1 \times 5 + 1 \times 1 \times 1$			$+1 \times -1 \times 1 + 1 \times -1 \times 1) = 1$	
$a(B_1) = \frac{1}{4}(1 \times 1 \times 5 + 1 \times -1 \times 1 + 1 \times 1 \times 1$			$+1 \times -1 \times 1) = 1$	
$a(B_2) = \frac{1}{4}(1 \times 1 \times 5 + 1 \times -1 \times 1 + 1 \times -1 \times 1 + 1 \times 1 \times 1)$ $= 1$				

Reading from the matrices the following representations can be deduced:

	E	C_2	$\sigma_v(xz)$	$\sigma_v(yz)$	
d_{z^2}	1	1	1	1	A_1
$d_{x^2-y^2}$	1	1	1	1	A_1
d_{xy}	1	1	-1	-1	A_2
d_{xz}	1	-1	1	-1	B_1
d_{yz}	1	-1	-1	1	B_2

Thus d_{z^2}, $d_{x^2-y^2}$, and d_{xz} belong to the same set of irreducible representations as the hydrogen orbitals.

4.4 D_{2h}, (y-axis joins bridge Hs, z-axis joins B atoms).

	E	$C_2(z)$	$C_2(y)$	$C_2(x)$	i	$\sigma_h(xy)$	$\sigma_h(xz)$	$\sigma_h(yz)$
(i)	2	0	2	0	0	2	0	2
(ii)	4	0	0	0	0	0	4	0
(iii)	6	-2*	0	0	0	0	$2^\dagger$	$2\ddagger$
(iv)	2	2	0	0	0	0	2	2

Reduction can be achieved by inspection of the character table

$\Gamma(\text{i}) = A_{1g} + B_{2u},$
$\Gamma(\text{ii}) = A_{1g} + B_{1u} + B_{3u} + B_{2g},$
$\Gamma(\text{iii}) = A_{1g} + B_{1u} + B_{2u} + B_{3u} + B_{2g} + B_{3g},$
$\Gamma(\text{iv}) = A_{1g} + B_{1u}.$

Cis-buta-1,3-diene belongs to the point group C_{2v}. The C atoms may be numbered as in the *trans* isomer. The matrices will be

$$
\begin{array}{c}
\Gamma \\
C_1 \\
C_4 \\
C_2 \\
C_3
\end{array}
\quad
E
\begin{bmatrix}
1 & 0 & 0 & 0 \\
0 & 1 & 0 & 0 \\
0 & 0 & 1 & 0 \\
0 & 0 & 0 & 1
\end{bmatrix}
\quad
C_2
\begin{bmatrix}
0 & -1 & 0 & 0 \\
-1 & 0 & 0 & 0 \\
0 & 0 & 0 & -1 \\
0 & 0 & -1 & 0
\end{bmatrix}
\quad
\sigma_v(xz)
\begin{bmatrix}
-1 & 0 & 0 & 0 \\
0 & -1 & 0 & 0 \\
0 & 0 & -1 & 0 \\
0 & 0 & 0 & -1
\end{bmatrix}
\quad
\sigma_v'(yz)
\begin{bmatrix}
0 & 1 & 0 & 0 \\
1 & 0 & 0 & 0 \\
0 & 0 & 0 & 1 \\
0 & 0 & 1 & 0
\end{bmatrix}
$$

The characters could be written down directly in two groups

$\Gamma(C_1, C_4)$ 2   0 -2 0 $(A_2 + B_2)$

$\Gamma(C_2, C_3)$ 2   0 -2 0 $(A_2 + B_2)$

---

* p_zs yield $+2$, p_xs yield -2, p_ys yield -2.
† p_zs yield $+2$, p_xs yield $+2$, p_ys yield -2.
‡ p_zs yield $+2$, p_xs yield -2, p_ys yield $+2$.

4.6 The allyl point group is C_{2v}.

	E	C_2	$\sigma_v(xz)$	$\sigma_v'(yz)$	
Γ(bonding orbital, $x = -\sqrt{2}\beta$)	1	-1	-1	1	B_2
Γ(nonbonding orbital, $x = 0$)	1	1	-1	-1	A_2
Γ(antibonding orbital, $x = \sqrt{2}\beta$)	1	-1	-1	1	B_2

5.1 Suppose the sp^2 hybrids to be

$$\left(\frac{1}{\sqrt{3}}s + \frac{1}{\sqrt{2}}p_y + \frac{1}{\sqrt{6}}p_x\right) \quad \text{and} \quad \left(\frac{1}{\sqrt{3}}s - \frac{1}{\sqrt{2}}p_y + \frac{1}{\sqrt{6}}p_x\right).$$

This leaves one-third of the s-orbital, two-thirds of the p_x orbital and all the p_z orbital to be used in the new hybrids. p_z must be divided equally and so must p_x. Thus the hybrids are

$$\left(\frac{1}{\sqrt{6}}s + \frac{1}{\sqrt{2}}p_z - \frac{1}{\sqrt{3}}p_x\right) \quad \text{and} \quad \left(\frac{1}{\sqrt{6}}s - \frac{1}{\sqrt{2}}p_z - \frac{1}{\sqrt{3}}p_x\right).$$

The ratio of the coefficients of p_z and p_x will determine the angle between the hybrid and the x-axis (α).

Thus $\quad \tan\alpha = \sqrt{\dfrac{3}{2}}$

and $\quad \alpha = 50°\ 46'$.

Thus the angle between the two new hybrids is $2\alpha = 101°\ 32'$.

Since the $s:p$ contribution is in the ratio $1:5$ these hybrids have been called sp^5 hybrids.

5.2 The character for a rotation through θ is

$$\chi = \frac{\sin(l+\tfrac{1}{2})\theta}{\sin\tfrac{1}{2}\theta}.$$

	C_2	C_3	C_4	
Thus $\Gamma(fs)$	-1	$+1$	-1	$(l = 3)$

(These values can also be found in Table 32.)

The symmetry operations for T_d are E, C_2, C_3, σ_d and S_4. Reflection in a mirror plane is equivalent to a C_2 rotation followed by inversion. Similarly S_4 is equivalent to $C_4 i$ in this point group. With respect to inversion f-orbitals are all antisymmetric.

Thus $\quad \chi_{\sigma_d}(fs) = \chi_{C_2}(f)\,\chi_i(f)$

$\qquad\qquad\quad = -1 \times -1 = +1$

and $\quad \chi_{S_4}(f) = \chi_{C_4}(f)\,\chi_i(f)$

$\qquad\qquad\quad = -1 \times -1 = +1.$

	E	C_3	C_2	σ_d	S_4
$\Gamma(f)$	7	1	-1	$+1$	$+1$

Irreducible representations for f-orbitals (T_d) are a_1, t_1 and t_2.

The oxygen lone pairs in IO_4^- or TeO_4^{2-} belong to irreducible representation $e + t_1 + t_2$.

Thus in principle at least six out of the seven f-orbitals could be used for π-bonding.

6.1 The orbitals are p_{+1}, p_0 and p_{-1}. Orbital functions for two electrons are:

	$\sum m_l = M_L$	Symmetric: antisymmetric
$\psi_{(i)} = p_{+1}(1)p_{+1}(2)$	$+2$	$+$
$\psi_{(ii)} = p_0(1)p_0(2)$	0	$+$
$\psi_{(iii)} = p_{-1}(1)p_{-1}(1)$	-2	$+$
$\psi_{(iv)} = (\sqrt{2})^{-1}\{p_{+1}(1)p_0(2) + p_0(1)p_{+1}(2)\}$	$+1$	$+$
$\psi_{(v)} = (\sqrt{2})^{-1}\{p_{+1}(1)p_0(2) - p_0(1)p_{+1}(2)\}$	$+1$	$-$
$\psi_{(vi)} = (\sqrt{2})^{-1}\{p_{+1}(1)p_{-1}(2) + p_{-1}(1)p_{+1}(2)\}$	0	$+$
$\psi_{(vii)} = (\sqrt{2})^{-1}\{p_{+1}(1)p_{-1}(2) - p_{-1}(1)p_{+1}(2)\}$	0	$-$
$\psi_{(viii)} = (\sqrt{2})^{-1}\{p_0(1)p_{-1}(2) + p_{-1}(1)p_0(2)\}$	-1	$+$
$\psi_{(ix)} = (\sqrt{2})^{-1}\{p_0(1)p_{-1}(2) - p_{-1}(1)p_0(2)\}$	-1	$-$

The three orbital functions (v), (vii) and (ix) must be associated with symmetrical spin functions, of which there are three (**1.4**, **1.6** and **1.7**) – giving nine wave functions in all. These comprise the ^3P state. The other six orbital functions can only be combined with the single antisymmetric spin function, equation **1.5**. Thus all other states will be singlets.

The total number of microstates will therefore be $9 + 6 = 15$.

(i), (iii), (iv) and (viii) are clearly associated with ^1D.

(ii) and (vi) both have $M_L = 0$, some linear combination of these orbital microstates will yield the orbital wave functions, one a component of ^1D; the other the orbital function for the ^1S state.

7.1 CO_2, *point group* $D_{\infty h}$

	E	C_x	C_2'	i	iC_x	(σ_h)	σ_v
Γ_{3N}	9	$3 + 6\cos\theta$	-1	-3	$-1 - 2\cos\theta$	1	3
Γ_{trans}	3	$1 + 2\cos\theta$	-1	-3	$-1 - 2\cos\theta$	1	1
Γ_{R_x, R_y} (only)	2	$2\cos\theta$	0	2	$2\cos\theta$	-2	0
Γ_{vib}	4	$2 + 2\cos\theta$	0	-2	$-2\cos\theta$	2	2

Hence $\Gamma_{vib} = A_{1g} + A_{2u} + E_{1u}$

$\qquad\quad = \Sigma_g^+ + \Sigma_u^+ + \Pi_u$.

There will be three fundamental frequencies of vibration (N.B. For a linear molecule (z-axis) only rotations about the x- and y-axes need be considered).

For infrared activity $\quad \Gamma(\mathscr{P}, \text{i.r.}) = A_{2u} + E_{1u}$.

For Raman activity $\quad\quad \Gamma(\mathscr{P}, \text{R}) = A_{1g} + E_{1g} + E_{2g}$.

The A_{1g} vibration will be Raman active and the A_{2u} and E_{1u} vibrations will be infrared active.

NO_3^-, *point group* D_{3h}

	E	σ_h	C_3	S_3	C_2^1	σ_v
Γ_{3N}	12	4	0	-2	-2	2
Γ_{trans}	3	1	0	-2	-1	1
Γ_{R_x, R_y, R_z}	3	-1	0	2	-1	-1
Γ_{vib}	6	4	0	-2	0	2

$\Gamma_{\text{vib}} = A_1' + A_2'' + 2E'$.

That is, there will be four fundamental frequencies of vibration.

For infrared activity $\quad (\mathscr{P}, \text{i.r.}) = A_2'' + E'$.

For Raman activity $\quad\quad (\mathscr{P}, \text{R}) = A_1' + E' + E''$.

Thus the A_1' vibration is Raman active, the A_2'' vibration is infrared active and the E' vibrations are active in both Raman and infrared spectra.

Index

Acetylene, symmetry of 63
Allyl, π molecular orbitals in 33,
  36
Ammonia
  bonding in 111
  determination of point group
    for 87
  matrices for H orbitals in 100
  symmetry of 59, 60, 64, 71
  vibrations in 185
Ammonium, NH_4^+, bonding in
  123
Arsenate, bonding in 125
Atom 11
  electronic states of 165
  helium 18
  valence state of 116
Atomic number 11
Atomic orbitals
  hybrid 53, 112
  hydrogen 14
  polarization of 51

Back-bonding 130
Benzene
  atomic coefficients in π-orbitals
    of 44
  determination of point group
    for 87
  electronic transitions in 162
  energies of π-orbitals of 43,
    136
  i.r. spectra of 189
  m.o. calculations for 43, 135
  Raman spectra of 189
  resonance in 52
  symmetry axes in 59
  symmetry of 62
  use of symmetry in m.o.
    calculations for 45
  vibronic coupling in u.v.
    spectra of 163
$B_6H_6^{2-}$, bonding in 147
$B_{12}H_{12}^{2-}$, symmetry of 85
Bond

bent 121
chemical 29
covalent 11
formation 53
ionic 55
localized 116, 118
three-centre 142
Bonding
  back, via π-orbitals 130
  in electron-deficient molecules
    141
  π, in organic molecules 130
  π, in transition metal
    complexes 129
  σ, in transition metal
    complexes 126–30
Borate, bonding in 124
Boride, B_6^{2-}, bonding in 145
Borohydride, BH_4^-, bonding in
  123
Boron–hydrogen bridge bonds
  143
Bravais lattices 196, 197
Buta-1,3-diene
  atomic coefficients in π-orbitals
    of 42
  bond order 42, 43
    in anion 43
    variation of 47
  charge distribution in π-orbitals
    of 42
  cis, electronic transitions in
    161
  energies of π-orbitals of 41
    using symmetry 131
  orbital symmetry conservation
    in isomerization of 156
  representations of π-orbitals of
    105
  trans, electronic transitions in
    161
  trans, π-orbitals of 104

C_∞, $C_{\infty v}$, continuous groups 78
C_n, C_{nh}, C_{nv} point groups 74

C_{ni} point groups 77
C_2 symmetry operation 58
C_3 symmetry operation 59
C_4 symmetry operation 61
C_5, C_6, C_7, C_8 symmetry
  operations 62
C_∞ symmetry operation 63
Calcium boride, bonding in 145
Carbon dioxide
  molecular orbitals in 48
  vibrational spectra of 187
Carbon monoxide, symmetry of
  63
Carbon $2p^2$ configuration
  microstates for 167
  terms for 168
Carbonate, bonding in 124
Carbonyl complexes, carbonyl
  stretching frequencies in 190,
  191
$C_{20}H_{20}$, symmetry of 85
Character 97
  for atomic displacements in
    molecules 185
  for p-orbitals for rotation
    through any angle 169
  simple method for determining
    101, 185
  tables 98, 211–34
    symbolism in 98
Chromium hexacarbonyl,
  bonding in 130
Chromophores 164
Class
  of crystals 194
  of symmetry operators 73
C.N.D.O. 49
Coefficients, atomic
  calculation of, in m.o.s 35, 39
  in degenerate m.o.s 39
  in π-orbitals of cyclic polyenes
    133
Complex conjugate 14
Complexes
  high-spin 128
  low-spin 128
  spin multiplicity of 176
  strong-field 128
  transition metal
    π-bonding in 129

σ-bonding in 126
  weak-field 128
Conjugation in organic molecules
  130
Configuration, electronic, in
  atoms 165, 166
Configuration interaction 27, 52
Con-rotatory reaction mechanism
  156
Continuous groups, C_∞, D_∞ 78
Coordinates
  angular 15
  Cartesian 15
  normal, of a molecule 181
  polar 15
Correlation
  charge 27
  diagrams for d^2 under O_h 176
  electron 26
  spin 27
Coupling
  j–j 167
  L–S 166
  Russell–Saunders 166
  spin–orbit 166, 177
    effect on d^1 configuration
      179
Crystal classes 194
Crystal field
  theory 165, 172
  O_h distortion of 171
  strong (O_h), effect on configura-
    tions, t_{2g}^x, e_g^x 173, 175
  weak, effect on atomic terms
    169–72
Crystals, structure of 194
Cubic point groups 83, 84
Cyclobutane
  symmetry of 61
  orbital symmetry conservation
    in isomerization of 156, 157
Cyclohexa-1,3-diene,
  isomerization of 157
Cyclo-octatetraenyl dianion,
  $C_8H_8^{2-}$, symmetry of 62
Cyclopentadienyl anion
  π-orbitals in 135
  symmetry of 62
  transition metal complexes of
    141

Cyclopropane
  delocalized bonds in   121
  symmetry of   59
Cyclopropenyl, m.o. calculations
  for   37–40

d^2, effect of O_h field on triplet
  terms of   177
Δ for d-orbital splitting in
  complexes   128
$D_\infty, D_{\infty v}$ continuous groups   78
$D_n, D_{nd} \cdot D_{nh}$ point groups   76
D_{ni} point group   77
De Broglie   12
Degeneracy   114
  of spin functions   24
  of wave functions   24
Degenerate orbitals, rotation of
  41
Determinant
  secular   32
    simplification of   109
  Slater   117
Diborane, bonding in   142
Dichloroethylene, *cis* and *trans*
  i.r. and Raman spectra of   188
  symmetry of   64, 69
Diels–Alder reaction   158
Direct product   98
Dis-rotatory reaction mechanism
  156
Dithionite, bonding in and
  configuration of   148–50
Dodecahedral point groups   84,
  85
Double groups   177
$d\tau$ 19

E, identity operation   58
e_g^2, wave functions for   173
Eigenfunction   13
Eigenvalue   13
Electron configuration
  in atoms   18
  of molecule, representation of
  160
Electron
  correlations   25
  energy of   12, 18
  mass   12

momentum   12
repulsion   18, 22, 24
  explicit consideration in m.o.
  calculations   49
velocity   12
Electronic states of atoms   165
Electronic transitions
  in carbonyl group   164
  in complexes   165
  in nitro group   165
  permitted (in atoms)   166
  in sodium   166
Electrons   11, 12
Element   11
Energy
  of atomic orbitals   26
  average   19
  kinetic   12
    of electron   12
  potential   12, 18
  promotion   116
  quantization of   12
  total   12
Equation, secular   32
Ethylene, bonding in
  m.o. model   119
  semi-localized bonds   120
  localized bonds   121
Excitation, electronic   159
Excited state, representations of
  162
Exclusion principle   18

Ferrocene
  bonding in   138–40
  symmetry of   65

Germanate, bonding in   125
Glide plane   200
Group
  Abelian   72
  continuous (C_∞, D_∞) 78
  double   177
  point   74
  space   200
  theory   72

h 12
Hamiltonian operator   13, 18,
  30, 108

Helium atom   18
  energy states of   21
  excited states of   22
  ground state of   20
  Hamiltonian operator for   21
Hermann–Maugin point-group
  notation   192
hexa-1,3,5-triene, isomerization
  of   157
Hückel   33
Hund's rule   24
Hybridization   111, 117
Hydrogen atom   14
  orbitals   14
  Schrödinger's equation   14
  wave functions   14
Hyperconjugation   122

i, inversion, symmetry operation
  69, 77
I, I_i, I_h point groups   84–5
Icosahedral point groups   84, 85
Improper rotations, S_n 66–8
Infra-red spectra
  overtones and combination
    bands   187–9
  selection rules for   186
Integral
  core   49
  Coulomb, α 33, 134
    variation of   48, 153
  Coulomb, J 22
  exchange   23
  overlap, S 33
  resonance, β 33, 134
    variation of   46, 143
Ions   11
Irreducible representations
  for angular momentum quan-
    tum numbers under O'
    (double group) symmetry
    179
  for atomic terms under strong
    O_h field   173, 175
  for molecular rotation (H_2O)
    183, 184
  for molecular translations
    (H_2O)   183, 184
  for orbitals in methane   102
  symbols for   98

for wave functions   102
Isotopes   11

Jahn–Teller effect   155
j–j coupling   167

K 23

Lattices
  Bravais   196, 197
  cubis   197
  monoclinic   201
  orthorhombic   198
  tetragonal   197
L.C.A.O.   29

Matrices
  for atomic displacements in
    water molecule   182
  irreducible   95
  multiplication of   209
  for orbitals in ammonia   93, 94
  reduction of   100, 101
  for orbitals in water   91
    reduction of   100
  reduction of   94, 99–102, 105
  similarity transformation   95
Matrix   89
  spur of   97
  square   95
  trace of   97
  unit   94
Methane
  bonding in   123
  ion, CH_4^+, bonding in and
    shape of   155
  molecular orbitals in flattened
    154–5
  representations for orbitals
    102
  sp^3 hybrids in   115
  symmetry of   60
Methyl, equivalent orbitals   122
Methylene group, vibrations of
  190
Microstates   167
  number possible   168
Molecular orbitals   29
  bond order of   37
  charge distribution in   37

effect of heteroatoms on 49
electron repulsion in 49
equivalent to atomic orbitals
104
representations of 105
Molecule
to determine point group of
86
shape of
CH_4^+ 154
H_2O 150
XeF_6 152
total energy of 31
with centre of symmetry, i.r.
and Raman spectra of 187
Molybdenum carbonyl com-
plexes, carbonyl stretching
frequencies in 191, 192
Molybdenum hexacarbonyl
bonding in 130
symmetry of 60
Monoclinic
lattices 201
space groups 201
Multiplicity 24, 167

Naphthalene, π molecular
orbitals in 137
Neutrons 11
Nitrate
bonding in 124
vibrational spectra of 187
Node 17
Normalization 14, 20, 36
of π-orbitals in cyclic polymers
133
Nucleus 11

O' double group, character table
for 178
O, O_h, O_i point groups 83, 84
Octahedral
point groups 83, 84
complexes 126, 128
Operator
electron repulsion 18, 23, 30
Hamiltonian 13, 18, 30, 108
mathematical 13
nuclear repulsion 30
Orbitals

antibonding 35, 141
steriochemical role of 153
atomic
energies of 26
hydrogen 14
see also Atomic orbitals
in B_6^{2-} unit 146, 147
bonding 35, 141
d 15
matrices for, in O_h 126, 127
polarization of 124
splitting of energy levels for,
in complexes 128
use in π-bonding 124, 148
degenerate 114
delocalized 111
in cyclopropane 121
in ethylene 119
in methyl 122
in diborane 142
electron distribution in 17
energies of 25
energy match of 54
equivalent 111, 117
in ferrocene 138
hybrid, sp, sp^2, sp^3 53, 112–15
interactions between 115
hybridization of 111
localized 111
in ethylene 120, 121
molecular 29
see also Molecular orbitals
nonbonding 35
orthogonal 16
overlap of 53
p 15
characters for, under general
rotation 169
π in benzene 136
conjugation of, in organic
molecules 130–38
in $C_5H_5^-$ 135
in cyclic polyenes, C_nH_n
132–6
in naphthalene 137
polarization of 51
rotation of 92
shapes of 15
Orbital symmetry match 55
Orthogonal functions 22

Orthogonality 16
  angle of 92, 169
Oxidation state, stabilization of
  high 129
  low 130
Oxyanions
  bonding in 124
  of transition metals, bonding
    in 129

P, P_i, P_h point groups 84, 85
Pauli exclusion principle 18
Pentaborane, B_5H_9
  bonding in 143
  symmetry of 61
Perbromate, bonding in 126
Perchlorate, bonding in 125
Phosphate, bonding in 125
$(PNCl_2)_3$, symmetry of 66
Point group
  determination of 86
  notation
    Herman–Maugin 192
    Schönflies 192
Point groups 74–85
  C_n, C_{nh}, C_{nv} 74
  C_{ni} 77
  character tables for 98
  cubic, O, O_h, O_i 83, 84
  D_n, D_{nd}, D_{nh} 76
  D_{ni} 77
  dodecahedral, I, I_i 84–5
  generation of 78–80
  icosahedral, I, I_i 84–5
  octahedral, O, O_h, O_i 83, 84
  P, P_i, P_h 84, 85
  S_n 76
  S_{ni} 77
  tetrahedral, T, T_d, T_i, T_h 81, 82
Polyenes, cyclic, energies of π-
  orbitals in 132–6
Protons 11

Quantum numbers
  angular momentum 15
  atomic, J, L, S 166
  in atoms 166
  electron spin 18
  magnetic 15
  principal 15

R, rotation through 360° in
  double group 177
Raman spectra
  overtones and combination
    bands in 187–9
  selection rules for 186
Reaction
  Diels–Alder 158
  mechanisms, orbital symmetry
    conservation in 155–7
Representation 90
  for atomic terms under O_h field
    171
  of degenerate orbitals 162
  of excited molecular states
    162
  of vibrational modes of a
    molecule 181
  wholly symmetric 108
Repulsion, electronic, in singlet
  and triplet states 168
Resonance 52, 118
  structures 116, 117
Rotations, improper, S_n 66–8

S_2, S_3, S_4 improper rotations 67
S_5, S_6, S_8 improper rotations 68
$S_n, S_{nh}, S_{ni}, S_{nv}$ point groups 76,
  77
Schönflies point group notation
  192
Schrödinger equation 13
  for simple harmonic oscillator
    186
Screw axis 198
Selection rules for
  electronic transitions 159
  i.r. spectra 186
  Raman spectra 186
Selenate, bonding in 125
Silicate, bonding in 125
Slater determinant 20
Sodium, electronic transitions in
  166
sp, sp^2, sp^3 hybridization 112–15
Space groups 200
  monoclinic 201
Spectroscopy
  i.r. 181
  Raman 181

effect of heteroatoms on 49
electron repulsion in 49
equivalent to atomic orbitals 104
representations of 105
Molecule
to determine point group of 86
shape of
CH_4^+ 154
H_2O 150
XeF_6 152
total energy of 31
with centre of symmetry, i.r. and Raman spectra of 187
Molybdenum carbonyl complexes, carbonyl stretching frequencies in 191, 192
Molybdenum hexacarbonyl bonding in 130
symmetry of 60
Monoclinic
lattices 201
space groups 201
Multiplicity 24, 167

Naphthalene, π molecular orbitals in 137
Neutrons 11
Nitrate
bonding in 124
vibrational spectra of 187
Node 17
Normalization 14, 20, 36
of π-orbitals in cyclic polymers 133
Nucleus 11

O' double group, character table for 178
O, O_h, O_i point groups 83, 84
Octahedral
point groups 83, 84
complexes 126, 128
Operator
electron repulsion 18, 23, 30
Hamiltonian 13, 18, 30, 108
mathematical 13
nuclear repulsion 30
Orbitals

antibonding 35, 141
steriochemical role of 153
atomic
energies of 26
hydrogen 14
see also Atomic orbitals
in B_6^{2-} unit 146, 147
bonding 35, 141
d 15
matrices for, in O_h 126, 127
polarization of 124
splitting of energy levels for, in complexes 128
use in π-bonding 124, 148
degenerate 114
delocalized 111
in cyclopropane 121
in ethylene 119
in methyl 122
in diborane 142
electron distribution in 17
energies of 25
energy match of 54
equivalent 111, 117
in ferrocene 138
hybrid, sp, sp^2, sp^3 53, 112–15
interactions between 115
hybridization of 111
localized 111
in ethylene 120, 121
molecular 29
see also Molecular orbitals
nonbonding 35
orthogonal 16
overlap of 53
p 15
characters for, under general rotation 169
π in benzene 136
conjugation of, in organic molecules 130–38
in $C_5H_5^-$ 135
in cyclic polyenes, C_nH_n 132–6
in naphthalene 137
polarization of 51
rotation of 92
shapes of 15
Orbital symmetry match 55
Orthogonal functions 22

Orthogonality   16
  angle of   92, 169
Oxidation state, stabilization of
  high   129
  low   130
Oxyanions
  bonding in   124
  of transition metals, bonding
    in   129

P, P_i, P_h point groups   84, 85
Pauli exclusion principle   18
Pentaborane, B_5H_9
  bonding in   143
  symmetry of   61
Perbromate, bonding in   126
Perchlorate, bonding in   125
Phosphate, bonding in   125
$(PNCl_2)_3$, symmetry of   66
Point group
  determination of   86
  notation
    Herman–Maugin   192
    Schönflies   192
Point groups   74–85
  C_n, C_{nh}, C_{nv} 74
  C_{ni} 77
  character tables for   98
  cubic, O, O_h, O_i 83, 84
  D_n, D_{nd}, D_{nh} 76
  D_{ni} 77
  dodecahedral, I, I_i 84–5
  generation of   78–80
  icosahedral, I, I_i 84–5
  octahedral, O, O_h, O_i 83, 84
  P, P_i, P_h 84, 85
  S_n 76
  S_{ni} 77
  tetrahedral, T, T_d, T_i, T_h 81, 82
Polyenes, cyclic, energies of π-
  orbitals in   132–6
Protons   11

Quantum numbers
  angular momentum   15
  atomic, J, L, S 166
  in atoms   166
  electron spin   18
  magnetic   15
  principal   15

R, rotation through 360° in
  double group   177
Raman spectra
  overtones and combination
    bands in   187–9
  selection rules for   186
Reaction
  Diels–Alder   158
  mechanisms, orbital symmetry
    conservation in   155–7
Representation   90
  for atomic terms under O_h field
    171
  of degenerate orbitals   162
  of excited molecular states
    162
  of vibrational modes of a
    molecule   181
  wholly symmetric   108
Repulsion, electronic, in singlet
  and triplet states   168
Resonance   52, 118
  structures   116, 117
Rotations, improper, S_n 66–8

S_2, S_3, S_4 improper rotations   67
S_5, S_6, S_8 improper rotations   68
$S_n, S_{nh}, S_{ni}, S_{nv}$ point groups   76,
  77
Schönflies point group notation
  192
Schrödinger equation   13
  for simple harmonic oscillator
    186
Screw axis   198
Selection rules for
  electronic transitions   159
  i.r. spectra   186
  Raman spectra   186
Selenate, bonding in   125
Silicate, bonding in   125
Slater determinant   20
Sodium, electronic transitions in
  166
sp, sp^2, sp^3 hybridization   112–15
Space groups   200
  monoclinic   201
Spectroscopy
  i.r.   181
  Raman   181

visible and u.v., selection rules
for 159
Spherical harmonic functions,
character for general rotation
of 170
Spin–orbit coupling 177
effect of, on d^1 configuration
179
Spiropentane
determination of point group
for 87
symmetry of 65
State
atomic 166
doublet 24
energies of 25
singlet 24
triplet 24
Structures
covalent 51
ionic 51, 117
nonbonded 117
resonance 116, 117
Sulphate, bonding in 125
Sulphur, determination of point
group for 87
Sulphur hexafluoride, symmetry
of 61, 83
Sulphur trioxide, symmetry of
64
Symmetry
axes, examples of 58–63
axis, principal 63
centre 69
conservation of, in chemical
reactions 155–7
element of 57
operation 57
effect of, on orbitals 88
identity, E 58
improper rotation, S_n 66–8
inversion, i 69, 77
reflection, σ 63–5, 76, 81, 83
rotation 58, 59
rotation, clockwise and
anticlockwise 59
translational 198
operators
classes of 73
commutation of 72

inverse 72
multiplication of 60, 62, 70,
71
representation by matrices
89
reduction of
$O \rightarrow D_3$ 172
$O \rightarrow D_4$ 175
$O_h \rightarrow C_{3v}$ 153
$T^d \rightarrow D_{2d} \rightarrow D_{4d}$ 154

$T, T_d, T_i (T_h)$ point groups 81–2
Terms
atomic 166, 167
for np^x, nd^x configurations
168
how to calculate numbers
of, from microstates 168
correlation of 177
symbol for 166
Tetrahedral
oxyanions, bonding in 124–6
point groups 81–2
Transitions
allowed 160
electronic, $\sigma \rightarrow \sigma^*$, $n \rightarrow \sigma^*$,
$\pi \rightarrow \pi^*$, $n \rightarrow \pi^*$ 159
in benzene 162
in buta-1,3-diene (cis and
trans) 161
in carbonyl groups 164
in chromophores 164
coupled with vibration 163
in nitro group 164
probability of 160
in water 161
forbidden 160
Rydberg 161
vibrational, probability of 186
Tropylium cation, $C_7H_7^+$,
symmetry of 62

Unit cell 194
possible rotational symmetry
operations of 195
symmetry restrictions on 195
Urea, crystal symmetry of 203

Valence bond method 52, 117
Valence state 116

Variation method   30, 31
Vectors, rotation of, through
   arbitrary angle   92
Vibrations, molecular   181
   representations for   181
Vibronic coupling   163

Water
   bonding in   109
   determination of point group
      for   87
   electronic transitions in   161
   energies of m.o.s of, as function
      of bond angle   150
   i.r. and Raman spectra of   187
   representations of normal
      coordinates of   181
   symmetry of   57
   vibrational modes of   184
   vibrations in   181
Wave equation   11
Wave function   12, 13, 29
   for ammonia H orbitals   103,
      104
   angular part of   20
   antisymmetric   19, 20
   degenerate   18, 24

determinant   19
electron correlation effects   27
electronic, representation for
   160
energy of   15
for helium   19
   excited state   22
for hydrogen atom   14, 17
irreducible representation of
   102
for m.o. in B_6^{2-} unit   146, 147
for methane H orbitals   103,
   104
molecular   51
nodes in   17
normalized   14
product   19
radial part of   17
spin   20
symmetric   20, 168
for vibrational state of a
   molecule   186
for water H orbitals   102
Wavelength   12

Xenon hexafluoride, bonding in,
   and shape of   152